George N Cross

Elementary Chemical Technics

A Handbook of Manipulation and Experimentation for Teachers of Limited

Experience

George N Cross

Elementary Chemical Technics
A Handbook of Manipulation and Experimentation for Teachers of Limited Experience

ISBN/EAN: 9783744646604

Printed in Europe, USA, Canada, Australia, Japan

Cover: Foto ©berggeist007 / pixelio.de

More available books at **www.hansebooks.com**

ELEMENTARY CHEMICAL TECHNICS

A HANDBOOK OF

MANIPULATION AND EXPERIMENTATION

FOR

TEACHERS OF LIMITED EXPERIENCE, AND IN SCHOOLS
WHERE CHEMISTRY MUST BE TAUGHT WITH
LIMITED APPLIANCES

GEORGE N. CROSS, A. M.,

PRINCIPAL OF THE ROBINSON FEMALE SEMINARY.

BOSTON
SILVER, ROGERS, & CO., PUBLISHERS
50 BROMFIELD STREET
1887

PREFACE.

THE office of this little book is expressed in its title. My faith that such a work may be of some service in the cause of science teaching, is founded in part upon a realization of what a valuable adjunct such a book would have been in the earlier years of my own teaching.

No attempt has been made to write a text-book of Chemistry. There are enough good works of that nature to meet the requirements of every school. But from a proper regard for brevity, and sometimes from a lack of appreciation of the teacher's needs on the part of authors, text-books fail to give that careful detail of direction which will insure success in experimentation.

In the chapter on the Metals I have made no mention of the more rare ones, considering only such as are likely to be worked with in the ordinary school courses.

The terms of both the English and Metric systems are used, for I find that the two systems are used about equally in schools, and the tables provided in text-books render the conversion from one system to the other an easy matter.

I wish to express my indebtedness to Prof. A. C.
BOYDEN of the Bridgewater Normal, and to Mr. H. W.
TYLER of the Massachusetts Institute of Technology,
who have read the manuscript and proof-sheets, and
given me valuable suggestions in my work.

GEORGE N. CROSS.

EXETER, N. H., May 25, 1887.

CONTENTS.

CHAPTER I.

PAGE

CONSTRUCTION AND EQUIPMENT OF LABORATORIES . . 1

CHAPTER II.

GLASS WORKING 18

CHAPTER III.

CONSTRUCTION OF APPARATUS 24

CHAPTER IV.

GENERAL MANIPULATIONS 35

CHAPTER V.

EXPERIMENTATION WITH HYDROGEN, OXYGEN, WATER, AND
AIR 44

CHAPTER VI.

EXPERIMENTATION WITH NITROGEN COMPOUNDS . . 55

CHAPTER VII.

FLUORINE, CHLORINE, BROMINE, AND IODINE, AND THEIR
COMPOUNDS 59

CHAPTER VIII.

EXPERIMENTATION WITH CARBON AND ITS COMPOUNDS   . 66

CHAPTER IX.

SULPHUR, PHOSPHORUS, ARSENIC, ANTIMONY, BORON, SILI-
CON, AND THEIR COMPOUNDS 76

CHAPTER X.

EXPERIMENTATION WITH THE METALS AND THEIR COM-
POUNDS 86

CHAPTER XI.

SPECTRUM ANALYSIS 100

CHAPTER XII.

ELECTRICITY IN CHEMICAL REACTIONS 104

ELEMENTARY CHEMICAL TECHNICS.

CHAPTER I.

CONSTRUCTION AND EQUIPMENT OF LABORATORIES.

1. An apartment devoted exclusively to the purpose is not only a great convenience, but an absolute necessity, in the successful teaching of Chemistry. Neatness, convenience, health of pupils and teachers, the risk of injury to physical and other kinds of apparatus, and to school furniture from the use of chemicals in ordinary schoolrooms, and an encouraging degree of success in experimentation, all demand such a room in every school where Chemistry is taught.

2. Selection of Rooms.— Rooms upon the highest floor of a school-building possess some advantages in the way of ventilation, without the danger of vitiating the air of apartments above. But ground-floors and basements oftener furnish the desired accommodations.

The apartment should be large, high-studded, and well lighted by several windows extending nearly to the ceiling. The windows should all open both at top and bottom for a quick change of air. Besides this it must have a large ventilator, with a strong draught, and adequate means of heating. An abundant supply of soft water is an absolute necessity in all chemical manipulations. Con-

1

venience in obtaining and drawing it off after using must
be studied. Equally important is a safe and economical
means of heating retorts, flasks, evaporating-dishes, etc.,
available for each pupil.

3. The two Laboratories. — It may be well to describe
minutely the furnishing and equipment of two small laboratories: the first constructed at trifling expense, but
of great practical service; the second somewhat more
elaborate and better equipped, but still within the means
of the majority of public and private schools. Both are
descriptions of laboratories actually built and successfully
used, and the quotations of figures are real ones, from the
carefully kept accounts of construction. In both, provi-
sions are made for the individual work of twelve pupils.
Of course the dimensions of apartments and the present
and prospective requirements of schools will modify these
calculations.

LABORATORY No. 1.

We will proceed on the supposition that the building
in which this laboratory is to be fitted up is supplied with
neither water nor illuminating-gas. These desiderata must
be supplied as well as possible.

4. Water Supply. — Obtain a sound, strong molasses-
hogshead (with iron hoops, if possible). Remove one
head, and set upon a strong platform of planks thirty-six
inches high, in a corner of the room, and near a window.
Get a tinman to conduct the water from the roof into the
hogshead. This can be done through the window, by removing a pane of glass, and inserting in its place a piece
of tin, perforated for 'the admission of the tin conductor.
From a point three inches below the top of the hogshead
carry out through the tin "pane" a waste-pipe of the same
size as the conductor. Obtain a piece of gas-pipe as long

as the distance from the hogshead to the end of the room. Have iron faucets fitted to the end and middle of this pipe, and by means of it carry water from the hogshead the length of the room at a height of three and one-half feet from the floor. If desirable, turn an iron faucet into the hogshead itself, and water can be drawn in three different parts of the room at once. The waste water can be poured into large earthen crocks provided for the purpose, collected at the close of the exercise and carried out.

5. **Heating.** — In lieu of Bunsen burners and gas, small glass lamps, burning alcohol, will be used for heating purposes.

6. Laboratory Table.

DIMENSIONS.

	FT.	IN.
Length	12	6
Width	4	4
Height	3	0
Height of partition $b\,b$ . .	1	6
" " shelf $a\,a$ . . .	0	9
Width of shelf a . . .	0	6
Height of shelves $c\,c\,c$ . .	2	0

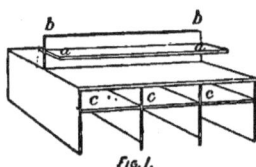

Fig. 1.

Fig. 1 represents a very convenient table adapted for twelve pupils. Each side is divided into three sections, each section being used by two pupils in common. Such an arrangement has many advantages. With every facility, abundant apparatus and chemicals, there is a great gain in providing for pupils in general chemistry to work in pairs. They are mutually helpful in the at first troublesome manipulations, and the saving in amount of apparatus and quantity of chemicals by such an arrangement is very considerable.

The figure will for the most part explain itself. It should be made of ⅞ pine stock, comparatively "clear." The top should be left unpainted. The remainder can be painted, or, better, covered with several coats of hard oil finish, or left in the natural condition, as taste and funds may direct. The shelf *a a*, and the space upon the table beneath, will be used for bottles and small pieces of apparatus. The partition *b b* should be carried down through to the floor, dividing the table into lateral halves. The shelves *c c c* below are for larger pieces of apparatus, and each pupil's private effects. Under the shelf in each section can be placed an earthen crock for the reception of waste water and other matter.

7. Storage Closet. — This should be made of the same well-seasoned stock as the table, with joints close fitting enough to keep out all dust. Convenient dimensions will be 8 feet in length, 7 feet in height, and 1 foot 5 inches deep. A central partition should divide the case into two sections. One side will be fitted up for the storage of chemicals. Leaving ample space in the bottom for the reception of large bottles of acids, let the upper part of the section be provided with shelves made each with three ledges, one above and back of another, like the successive steps in a flight of stairs. These ledges, about three inches in height, will be a great convenience in the arrangement of bottles of different sizes, placing the largest upon the highest ledges.

The other section will receive the glassware and apparatus, and can be made convenient by means of plain, deep shelves at suitable distances from each other.

8. Lecture Table. — There should be sufficient space in the room chosen for a laboratory, or an adjoining one, for the recitations on the subject, and the lecture table should be conveniently placed in this room. Such a table is indispensable for the performance of general class experiments, not profitably to be given to the members of a class, for illustrations in the course of a recitation, and as furnishing a deep tank necessary in the transferring of gases from one vessel to another.

For illustration of such a table see Fig. 3. The table for Laboratory No. 1 is of the same general construction as the one described in Section 14, but may be made of cheaper materials and painted in a less expensive manner. Considerable expense may be saved by building, instead of the drawers, plain lockers, and the convenience of these may be enhanced by putting a horizontal shelf through the middle of each locker.

9. Hood. — A close box or hood, within which all operations involving the generation of poisonous gases can be carried on, should be a part of the equipment of every laboratory. This box should be well lighted, constructed so tight that no gas can escape into the room, and arranged to conduct the gases generated either into a ventilator-box, with an unfailing upward draught, or directly to the open air through a window. In Section 16 is described a very perfect but somewhat expensive hood. An excellent substitute is the following:

Make a box-frame of $\frac{7}{8}$ pine stock 18 inches deep, of such dimensions as to be set within the frame and fit up closely against the sash of the window least used in the laboratory. Purchase a glazed sash of the size of the inside dimensions of the box-frame and arrange it to slide up and down in one side of the frame like a window. With hooks on the box and screw-eyes in the lower sash of the window, clamp the box closely to the window so that the lower window-sash shall form one side, and the new sliding sash the other side of a tight box. Fit into the top of the box an elbow-joint of small stove-pipe, and pass this pipe out through the upper sash by removing one of the panes of glass and substituting a piece of tin perforated for the passage of the pipe. For ventilation of the box bore several holes in the lower rail of the window-sash, and, to insure an upward draught, a little more than half-way up on one of the wooden sides of the box place a lamp-bracket, and set a lighted kerosene lamp in the bracket whenever the hood is to be used. The cost will be about $2.50.

10. Equipments.

APPARATUS AND GLASSWARE.

Alcohol lamps. (See section 45.)
Aspirators. (See section 30.)
Bags for gases. (See section 32.)
1 dozen beakers, 2½ ounces, at .15 .   . . $1.80
Bell-jars, tubulated. (See section 38.)
Mouth blowpipes. (See section 40.)
Gas-bottles (horseradish bottles, etc., from home).
3 dozen 4-oz. bottles, with ground stoppers, at $1.35   4.05
1 dozen one-quart fruit-jars 1.50
Book of chemical labels 50
6 clamps for test-tubes. (See section 30.)
6 earthen crocks 72
10 dozen assorted corks 1.20
Cover glasses. (Cut up pieces of broken window-
  panes into squares 4 × 4 inches.)
Set of cork-borers. (See section 41.)
2 round files 40
1 shoeknife for cutting corks 35
1 flat file 35
6 Hessian crucibles, 4½-inch 60
6 deflagrating spoons. (See section 42.)
6 porcelain evaporating-dishes, 3-inch   .   . . 1.50
1 pack of filter-paper, 6-inch 40
4 flasks, flat bottomed, 8-ounce 1.85
12 flasks, flat bottomed, 4-ounce .   .   .   1.20
6 glass funnels, 3-inch 90
2 lbs. glass-tubing (assorted numbers) .   .   1.00
2 graduates, 4-ounce 80
6 large nappies (for pneumatic pans) .   ·   .   2.10
Generators for hydrogen, etc. (See section 31.)
        Amount carried forward .   .   . $21.22

Amount brought forward . . . § 21.22
3 Wedgewood mortars, 3-inch 2.40
Platinum wire, ¼ pwt. 30
" foil, ¼ pwt. 30
2 potash bulbs (one bulb) 80
1 iron retort 3.00
24 feet rubber-tubing, ₅/₁₆-inch [1] 3.60
1 pair of scales, platform 5.00
6 retort stands. (See section 35.)
3 dozen test-tubes, 6-inch 1.08
6 thistle-tubes 72
6 test-tube brushes. (Tie pieces of fine sponge to a stout wire.)
3 U-tubes, 4-inch 75
Wash-bottles. (See section 43.)
Wire gauze, 1 square foot 50
Lead-pan. (Make it from sheet-lead, as directed for making sand-bath. See section 30.)

Total . . . . § 39.67

CHEMICALS.

Alcohol, 2 quarts. (Price variable.) . . . $ 1.20
Alum, 8 ounces 05
Aqua ammonia, 1 pound 20
Ammonium chloride, 1 pound 25
Ammonium nitrate, 1 pound 40
Antimony, metal, 4 ounces 10
Arsenious acid, 2 ounces 10
Bone-black, 1 pound 30
Calcium chloride 40
" sulphate, 1 pound 10
Camphor, ¼ pound 25
Carbon disulphide, 1 pound 50
Chlorhydric acid, 5 pounds 35
Cochineal, 2 ounces 10
Copper-filings, ¼ pound 25
" oxide, 1 ounce 20
" sulphate, 1 pound 12

Amount carried forward . . . $ 4.87

[1] Some tubing of a smaller size may be used for connection.

Amount brought forward . . . $4.87

Ether, 1 pint	.80
Fluor-spar, 1 pound	.15
Iron sulphate, 1 pound	.05
Lead acetate, 2 ounces	.10
" oxide, 8 ounces	.10
Litmus, 1 ounce	.15
Magnesium ribbon, 1 pwt.	.25
Manganese, black oxide, 2 pounds . . .	.30
Nitric acid, 5 pounds	.75
Nut-galls, 2 ounces	.10
Oxygen mixture, 3 pounds	1.20
Phosphorus, 1 ounce	.25
Potassium, metallic, ¼ ounce	1.00
" bichromate, 1 pound	.25
" bromide, 4 ounces	.40
" carbonate, 1 pound	.15
" iodide, 2 ounces	.80
" hydrate, 1 pound	.75
" chlorate, 2 pounds	.60
" cyanide, 2 ounces	.10
" ferrocyanide, ¼ pound	.25
" nitrate, 1 pound	.20
" permanganate, 1 ounce	.20
" tartrate, 2 ounces	.10
Silver nitrate, 1 ounce	.90
Sodium, metallic, ¼ ounce	.50
" acetate, ½ pound	.30
" bicarbonate, 1 pound	.12
" carbonate, 1 pound	.06
" hydrate	.75
" chloride, coarse, 5-pound box . . .	.10
" " fine, 3 pounds	.09
" nitrate, 1 pound	.25
" biborate, ½ pound	.10
" silicate, 1 pint	.15
" sulphate, 1 pound	.15
Strontium nitrate, 4 ounces	.10
Sulphur-roll, 1 pound	.06

Amount carried forward . . . $17.50

Amount brought forward   .   .   . $17.50
Sulphur-flowers, 1 pound 10
Sulphuric acid, 10 pounds 25
Turpentine, spirits of, ¼ pint 30
Zinc chloride, 1 ounce 25
Zinc, granulated.   (See section 58.)
    Total $18.40

Almost every manufactory of iron, woollen, or cotton goods uses sulphuric acid, where it can be purchased (as an accommodation) at the price quoted. Manufactories of woollen fabrics use chlorhydric and nitric acid. These acids, commercially pure, can often be obtained at such places much cheaper than of dealers in chemicals. Many of the chemicals will keep, unimpaired in quality, for an indefinite period. It is much cheaper to purchase such in bulk, and it is so directed in the above list. Keep all such chemicals in one and two quart fruit-jars. They are cheaper than regular salt-bottles, and, having wide mouths and tight covers, are equally good. Thus preserved the supply of many substances given will be sufficient to last for years.

11. Cost of such a laboratory, fitted up for a class of twelve pupils in general chemistry, is:

Hogshead and water-pipe $3.50
Large table 25.00
Lecture table 16.00
Tank 5.30
Storage closet 20.00
Hood 2.50
Apparatus and glassware 39.67
Chemicals 18.40
    Total $130.37

12. Laboratory No. 2.—This provides for the same number of pupils as the first described, but is fitted up in a

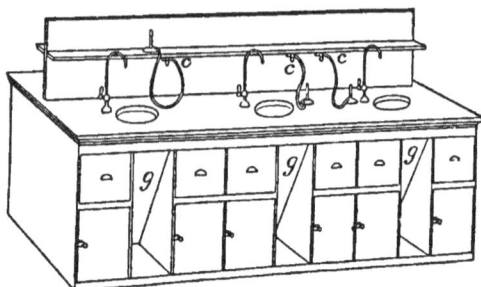

FIG. 2.

more complete and expensive manner, for work both in general chemistry and qualitative analysis.

It presupposes that the school building is supplied with gas and water, and provides only for bringing both into the room, and carrying the waste-water out.

13. Working Table. — Fig. 2 represents the table for the use of the pupils. It is made of white-wood, and stained, or of hard wood, and filled, and is of the same dimensions as given in Fig. 1. Each pupil has a drawer, 6 inches deep, for private effects, and a locker, 22 inches high, for larger pieces of apparatus.

In the enclosed spaces $g\,g\,g$, 1 foot in width, necessarily created by the set bowls, are placed earthen crocks to receive solid matter that might clog the waste-pipes.

The backs of these little recesses should be readily removable in order to give access to the interior of the table, in case the plumbing gets out of order.

On each side are three set bowls, nine inches in diameter, without stoppers, each conveniently placed for the use of two pupils. At $c\,c\,c$ are gas-cocks furnished with straight tubes for supplying gas to the Bunsen burners by means of $\frac{3}{8}$-inch rubber tubing.

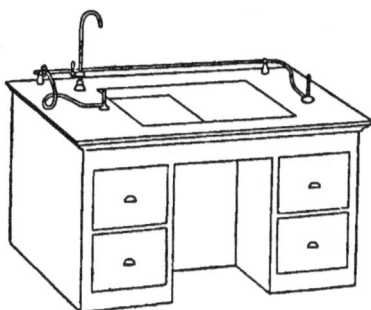

FIG. 3.

DIMENSIONS.	FT.	IN.
Length	8	0
Width	3	10
Height	3	0

14. Lecture Table. — Fig. 3 represents the lecture table. It is of white-wood, or hard wood stained or filled, and varnished. The top should only be treated to several coats of shellac.

At *a* is a lead-lined water-tank, 30 × 18 inches in area, and 24 inches deep at one end. The other end should be but 3 inches deep for a distance of 12 inches. By this means bell-jars can be filled with water, inverted, and stand in shallow water. This is a much better arrangement than soldering cleats upon the sides of the tank and putting in slate shelves. Such shelves are pretty sure to break down at an unfortunate time.

The table is supplied with water-cock, waste-pipe from the tank, and two gas-cocks for supplying fuel to two Bunsen burners.

15. Storage Closet. — The plan is the same as described in section 7, but the closet is built of better stock, ornamented with a cornice at the top, painted, has four glass

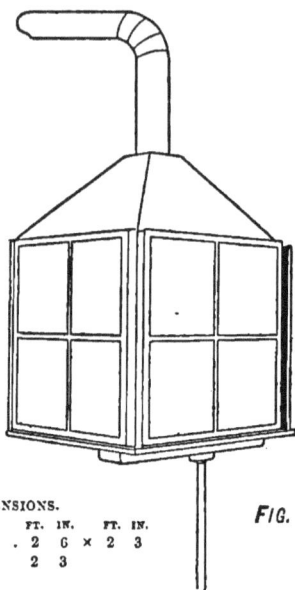

DIMENSIONS.

		FT.	IN.	FT.	IN.
Base	. . .	2	6	× 2	3
Height	. .	2	3		

FIG. 4.

doors instead of wooden ones, as in the cheaper closet, and is of larger dimensions, being 11 feet in length, 7 feet in height, and 1 foot 5 inches deep.

16. Hood. — Fig. 4 represents a convenient hood. It consists of a chamber with bottom and two sides of wood. The remaining sides are window-sashes with four 11 × 11 inch lights. One of the sashes is made to slide up and down as in an ordinary window. A dome-like roof of zinc is securely fitted upon the top of the chamber. The four sloping sides of this roof are narrowed until they reach a six-inch collar, over which is slipped a length of funnel which passes into a ventilator-box. The chamber is fastened, by means of large brackets, to the side of the room at a convenient height. Inside the chamber, upon one of the wooden sides, screw up a lamp-bracket to hold a large kerosene lamp. This should be kept constantly burning, while the hood is in use, to assure an ascending current. Remember that an upward draught cannot be secured without a corresponding inlet for cold air, and raise the sash a very little. If a ventilator-box is not available, carry the pipe of the hood out through a window, as directed in section 9.

17. Plumbing. — The water-pipes should be so placed that every foot of them can be easily reached. Let the discharge pipes from the bowls be carried into a drain-tile main, and the latter must be securely trapped. Provide for shutting off the water supply and draining the pipes in cold weather.

18. Equipments.

APPARATUS AND GLASSWARE.

Beakers, 2 dozen 2½-ounce	$3.60
8 Bunsen burners, with ring	4.00
6 mouth blowpipes	1.50
1 hot-blast blowpipe, for working glass	3.00
1 book chemical tables	.50
6 dozen glass bottles, 4-ounce, ground stoppers	8.10
1 nest Hessian crucibles	.30
6 earthen crocks	.72
1 set of brass cork-borers	2.50
6 test-tube brushes	.72
10 dozen corks, assorted sizes	1.20
Cover glasses. (See preceding list of apparatus.)	
6 deflagrating spoons	1.20
6 evaporating dishes, 3-inch	1.50
1 pack of filter-paper, 6-inch	.40
3 files, round, flat, three-cornered	.60
2 dozen 4-ounce flasks	2.40
Amount carried forward	$32.14

Amount brought forward . . . § 32.14
4 dozen 8-ounce flasks 85
6 funnels, 3-inch 90
2 pounds glass-tubing (assorted numbers) . . 1.00
6 generators. (See section 31.)
1 graduate, 4-ounce 40
1   " English and Metric 90
2 gas-bags, 55 gallons 25.00
1 dozen gas-jars (use fruit-jars) 1.50
1 dozen litmus-papers 40
6 Wedgewood mortars, 3-inch 2.40
6 nappies (large for pneumatic pans) . . . 2.10
40 feet rubber-tubing, $\frac{3}{16}$ 6.00
Chemical scales 12.00
6 iron retort-stands, 3 rings 5.00
1 iron retort for making oxygen . . . 3.00
1 chemical thermometer 2.25
6 dozen test-tubes, 6-inch 2.16
3 dozen " " 4-inch 1.20
Test-tube holders. (See section 36.)
6 thistle-tubes 72
3 U-tubes, 4-inch 75
3 chloride of calcium tubes 90
Tube for hydrogen tone 50
Wash-bottles. (See section 43.)
Brass wire gauze 50
Iron   "   "   (use mosquito-netting).
    Total . . . . § 102.67

CHEMICALS.

Alcohol, 1 quart . . . . § 0.60
Arsenious acid, 2 ounces 10
Alum, ¼ pound 05
Ammonia, 5-pint bottle 60
Ammonium carbonate, 1 pound 30
    " chloride, " " 25
    " nitrate, crystals, 1 pound 40
Aniline, red, ¼ ounce 30
Antimony (metallic), ¼ pound 20
    " tartrate (Tartar emetic), 1 ounce . . .10
    Amount carried forward . . . $ 2.90

Amount brought forward . . . § 2.90

Barium hydrate, 1 ounce	.25
Barium sulphate, 1 "	.10
Benzole, 1 pint	.30
Bone-black, 1 pound	.30
Borax, 1 pound	.20
Bromine, 1 ounce	.25
Calcium chloride, 1 pound	.40
" fluoride, 1 "	.15
" sulphate, 1 "	.10
Camphor, ½ pound	.25
Carbolic acid, 1 ounce	.15
Carbon disulphide, 1 pound	.50
Cochineal, 2 ounces	.10
Copper trimmings, 1 pound . . .	.50
" wire gauze	.40
Copper oxide, 1 ounce	.20
" sulphate, 2 pounds	.24
Chlorhydric acid, 5 pounds	.35
Ether, sulphuric, 1 pint	.80
Nut-galls, 1 ounce	.05
Gun-cotton, ¼ ounce	.15
Iodine, 1 ounce	.50
Iron sulphate, 1 pound	.05
Lead acetate, 2 ounces	.10
" carbonate, 1 pound	.18
" nitrate, 2 ounces	.10
" oxide, ½ pound	.10
Lithium carbonate, ¼ ounce	.25
Litmus, 1 ounce	.10
Magnesium, 1 pwt. (ribbon)	.25
" sulphate, 1 pound	.10
Manganese, black oxide, 3 pounds . . .	.45
Mercury, 2 pounds	1.50
" red oxide, 2 ounces	.30
Nitric acid, 5 pounds	.75
Nickel sulphate and ammonia, 2 ounces . .	.40
Oxygen mixture, 5 pounds	2.00
Phosphorus, 1 ounce	.25

Amount carried forward . . . § 21.62

Amount brought forward . . .	$ 16.02
Platinum foil, 1 pwt.	.60
Platinum wire, 1 pwt.	.60
Potassium, ¼ ounce	1.00
" bromide, 2 ounces	.20
" bicarbonate, 1 pound	.40
" bichromate, 2 pounds . . .	.50
" carbonate, 1 pound	.20
" hydrate, 1 pound	.75
" chlorate, 3 pounds	.90
" cyanide, ¼ pound	.20
" ferrocyanide, ½ pound	.25
" iodide, 1 ounce	.40
" nitrate, 1 pound	.20
Potassium permanganate, 1 ounce . . .	.20
Silver nitrate, 1 ounce	.90
Sodium, metallic, ¼ ounce	.50
" acetate, ¼ pound	.30
" bicarbonate, 1 pound	.12
" carbonate, 1 pound	.06
" hydrate, 1 pound	.75
" chloride, 5-pound box	.10
" " (coarse fine) 3 pounds . .	.09
" nitrate, 1 pound	.25
" silicate, 1 pint	.15
" sulphate, 1 pound	.15
Strontium nitrate, ¼ pound	.20
Sulphur-roll, 1 pound	.06
" flowers, 1 pound	.10
Sulphuric acid, 10 pounds	.25
Tinfoil (pure), ¼ pound	.35
Turpentine, 1 pint	.00
Zinc, chloride, 1 ounce	.25
Zinc, granulated, 4 pounds	1.20
Total	$ 28.80

19. Cost of Laboratory No. 2.

Plumbing	45.00
Laboratory table	40.00
Amount carried forward . . .	$ 85.00

Amount brought forward . . .	$85.00
Lecture	. 23.00
Tank	5.30
Storage closet	. 45.00
Hood	14.00
Painting	. 20.00
Apparatus and glassware	102.07
Chemicals	. 28.80
Total	. $323.77

The list of chemicals is a general one, selected with a view to the use of no particular text-book. Of course the list can be modified to adapt it to any series of experiments a teacher may choose to have demonstrated. The prices of apparatus and chemicals quoted are many of them list prices. If purchases are made in considerable quantities, as is always advisable, from 10 to 15 per cent may be deducted in the estimates of the amount of the last two items.

20. Labelling Chemicals. — All bottles and packages of chemicals should be labelled before they are placed in the storage closet. Books of printed labels, with structural formulas of the substance named, can be purchased, and are exceedingly convenient. In lieu of these, gummed labels $1\frac{1}{4} \times 1$ inch may be purchased in small boxes. The names should be written accurately in ink.

CHAPTER II

21. A Supply of Soft Glass Tubing should be purchased by the pound and kept constantly on hand.

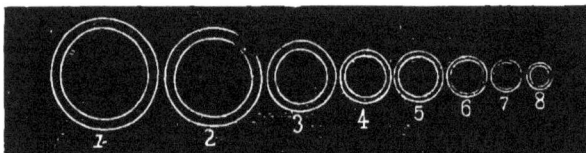

Fig. 5.

The sizes should run from one to eight. Sizes 3 and 4 will be used more than others. Rubber tubing of the same calibre can be readily obtained, and tight but easily adjusted joints thus made. A small quantity of hard glass tubing, of sizes 2, 3, and 4, will also be required. Small tubing can be broken squarely at any point by scratching the glass at that point with a three-cornered file, and applying pressure against the opposite side of the tube from the scratch. The sharp broken edges should be removed by softening the end in the gas-flame or by grinding the ends of the tube in the manner described in Section 23.

22. Boring Glass. — Holes large or small can be bored through the hardest glass, without any danger of cracking or shivering the glass, as follows:

Dissolve gum-camphor in spirits of turpentine until a nearly saturated solution is obtained. Provide a number

18

of round files (new) of different sizes. Break off the tip
of each file so as to make a sharp, ragged end. Select the
smallest file. Dip the end in the turpentine solution, and,
holding the file at an angle of about 45° with the surface
of the glass, at the same time press the ragged tip against
the glass and move the file rapidly to and fro. A depression will immediately be made in the glass, and its thickness soon cut through. The teeth of the file can then be
used. Keep the files constantly wet with the solution.
Enlarge the orifice with files of successive sizes. Use files
but a little smaller than the orifice itself.

23. Grinding Glass. — Large surfaces of glass can be cut
away and ground very rapidly, and without any danger of
breaking, by using flat files constantly wet in the turpentine solution. If accurate grinding of plane surfaces is
required, another device is necessary. Bell-jars may be
made to fit the pump-plate, or pneumatic jars squared off
for use with cover-glasses, on a lathe. Prepare a disk of
hard wood 1¼ inches in thickness, and with a diameter considerably larger than that of the largest jar to be ground.
Melt together pitch, tallow, and glue in the proportion of
10 parts of the first, 2 of the second, and 5 of the last, and
stir into the fluid just before it cools (not too soon, or it
will settle irregularly through the mass) a large quantity
of medium fine emery. Coat the face of the disk thickly
with this mixture while it is hot enough to spread easily.
In a few hours the disk will have a good grinding surface.
Attach this improvised emery-wheel to the shaft of the
lathe, or even upon a common whirling-table. Keep the
surface wet with the turpentine solution. Hold the jar
firmly and squarely against the revolving disk.

24. Cutting Glass. — Plane surfaces of glass can be cut in
straight lines or curves; large tubing, necks and bottoms
of bottles cut off very easily and rapidly by taking advan-

tage of the expansion of glass when heated. The disks of
hardened steel made for cutting glass often work well for
a short time, but are liable to give out suddenly. Good
diamonds are expensive, easily broken, will cut well only
in straight lines, and always leave a sharp cutting-edge.
None of the objections are to be met with in using heat.
The tools are very simple and inexpensive. Procure two
rods of soft iron (not steel) from $\frac{3}{8}$ to $\frac{1}{2}$ inch in diameter,
one 12, the other 15 inches in length. The best possible
material is the soft iron rod, used to fasten together the
parts of a stove, obtainable at any stove store. The shorter
one should be perfectly straight. Bend the longer one in
a somewhat sharp compound curve. Insert them in wooden
handles which can be purchased at a hardware store. To
cut a plane surface, as a window-pane, heat the straight
rod, if the glass is quite thick ($\frac{1}{4}$ inch or more), to a
red heat; if thinner, not so hot. Then with a three-cornered file, wet with the turpentine solution, cut a deep
groove in the edge of the glass and place the glass upon
the even surface of a table. Place the heated rod flat
upon the glass, with the end $\frac{1}{16}$ of an inch below the file-
mark. Now draw the rod downward in the direction in
which you wish the cut to run, and the crack will follow
if you keep the rod about $\frac{1}{4}$ of an inch ahead of it. If the
rod is very hot, be careful that the crack does not quite
catch up. If it does, it may dart on and take a direction
of its own. With a camel's-hair brush dipped in black
paint, or with pen and ink, mark out any design in circles
or scallops, curves or straight lines. A little practice will
enable one to cut the glass by these lines with perfect
accuracy, and the edges will be much smoother than when
cut with a diamond.

25. **To cut open Large Tubing**, bottles, etc., heat the
curved rod. With the file, wet as before, cut a groove from

¼ to ½ inch in length, according to the size of the tubing or bottle, and nearly through the thickness of the glass. Wipe off the turpentine solution from the vicinity of the groove. Place the heated rod first against the end of the groove in the position *a* in the figure, and wait until the glass is heard to crack. Then press it against the other end of the groove in the position *b*, and the crack will start and follow the rod round the bottle in the direction indicated by the arrow.

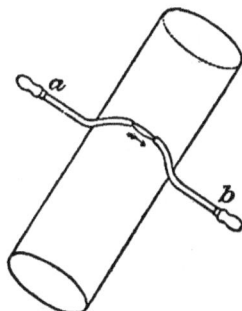

FIG. 6.

26. To bend Tubing. — To bend large tubing pack it full of fine sand, and heat it in a charcoal or bituminous-coal fire. Get a blacksmith to accommodate you with his forge fire.

For bending small tubing do not use a Bunsen burner or alcohol lamp. The best possible flame is obtained by removing the tube of a Bunsen burner and screwing on in its place a fish-tail burner. Hold the tube lengthwise in the flame, constantly turning it. The moment when it is soft enough to bend is judged better by the feeling, as it turns in the fingers, than by the eye. To prevent its flattening, bend it about 30° in one direction, then remove it from the flame and bend it slowly to the required angle in exactly the opposite direction. If a gas-flame cannot be had, use the flame of a common kerosene lamp, or, better, lantern. The deposit of carbon is easily wiped off with a piece of flannel while the tube is warm.

27. To Etch Graduations upon Eudiometers, etc. — Be sure the tube to be marked is perfectly clean. If not, wash it with strong soap-suds. Then cover it with a thin coat of paraffine. This may be done by rubbing the heated tube with a paraffine candle. The thinner the coat, to completely cover the surface, the better. The graduations may be of inches or centimeters. If the latter is desired, excellent finely divided scales may be obtained of any agent of the Metric Bureau. Place the tube and yard-stick, or whatever scale is used, together horizontally upon a table, blocking up the latter so that the surfaces of the tube and scale are exactly level. Then with a carpenter's try-square, set squarely against the side of the yard-stick, you can make the markings on the glass exactly correspond with the divisions of the scale. Mark completely through the wax with a sharp-pointed needle. Place figures at the inch or centimeter divisions. In a leaden pan, or a piece of sheet-lead folded into a sort of pan 3 or 4 inches square, mix up a thin paste of calcium flouride and strong sulphuric acid (a table-spoonful of the former will be enough), and arrange the tube, with marked side down, horizontally over the lead pan. Fold a large piece of stout paper over the tube, so as to completely shut in the tube and pan. Gently warm the pan for 3 or 4 minutes. The heat must be very moderate, so as not to melt the wax. Remove the heat, and let the apparatus stand several hours, — all night, if convenient. Do this work under a hood or in a strong draught, for the fumes are *very corro-*

sive to the lungs. Finally, remove the wax and clean the tube with spirits of turpentine.

28. To Close Tubes. — Mark off upon a long piece of tubing the length of the required tube. Heat the tubing at that point, constantly revolving it in the flame. When it is soft enough, suddenly, and in a direct line, draw it out till the bore is closed and the tube parts. Then heat the end to be closed again till the glass becomes plastic, and with the heated handle of an iron spoon press the tapering end back, and round it into shape. If the end to be closed is not attached to a long rod, first heat the end till it is soft, then while it is still in the flame press another piece of soft tubing against it till the two are welded. The second piece will then constitute a handle by which the tube can be drawn out and then closed.

CHAPTER III

29. SOME pieces of apparatus it is highest economy to purchase, made of the best materials and in the best manner. On the other hand, the list of essentials for the outfit of a chemical laboratory that can be made by the ingenious teacher or pupil, or furnished from home to the wonderful lessening of bills of expenditure, is a very long one. Let it be a rule that pieces of apparatus made for a special purpose, after using, shall be put away clean, in a case, and sacredly kept for that purpose. It is poorest economy not to furnish materials enough to obviate the make-shift necessity of pulling a piece of apparatus apart to obtain the tubing and other portions to supply a present need. Let everything be prepared in advance of the time for its use. A cork bored or fitted to a bottle in a hurry is almost sure to leak.

Every piece of apparatus described in this chapter will meet the requirements for which it is intended, if well made.

FIG. 6.

30. Aspirators. — Fig. 6 is an aspirator made from a two-quart pickle-jar. The larger the bottle the better for many purposes. a is an inlet tube of No. 4 tubing 6 inches long. At b a hole $\frac{1}{4}$ of an inch in diameter is drilled, and the orifice closed with a cork.

To use it, fill the jar with water by opening b, and

24

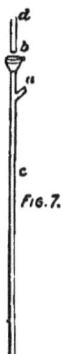

Fig. 7.

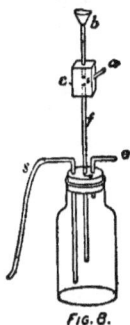

FIG. 8.

holding the jar under water till full. Then connect *a* by rubber tubing with the vessel through which a draught is desired, and let the water run into a vessel prepared for its reception.

A small aspirator of this kind should be prepared, of a half-pint capacity, or less, for use in some experiments, as withdrawing gas from candle-flame (Section 129, *b*). In experiments requiring a steady and uninterrupted draught for some time this aspirator is unsatisfactory. For such purposes make the following:

Fig. 7 is a tin tube $\frac{3}{8}$ of an inch in diameter and 36 inches long. *a* is an inlet tube a trifle smaller, 3 inches long, inserted 8 inches below the tunnel *b*. *d* is a rubber supply pipe connected with a water-cock.

To use, connect *a* with the vessel through which the draught is to be made, and turn on the water at *d*. The dimensions given are not essential, but the tube must be small enough for the supply of water to keep the neck of the tunnel *b* constantly full, and *a* should be of the right size to slip the rubber tubing over.

Fig. 8 has an arrangement for making a draught by using
a or of obtaining an air-blast by connecting with *e*.   *c* is
a piece of sheet-cork $2\frac{1}{2}$ by 2 by $2\frac{1}{4}$ inches, dipped in
melted paraffine to fill its pores.   *b* is a glass tunnel and
f No. 3 glass tubing.   *s* is a siphon for withdrawing the
water.   Regulate the length of *s*, so that it shall not
remove the water certainly any faster than it comes in.
If a water supply is not at hand, *a* can be used to good
advantage, and even *e* to some purpose, by steadily pouring
water into *b* from a large pitcher or nosed tin-pail.

Fig. 9 represents a convenient piece of apparatus for
obtaining a draught through a liquid.   *a* is a piece of glass

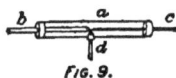

FIG. 9.

tubing one inch in diameter, and 8 or 9
inches long.   The top of an argand
lamp-chimney is suitable.   The ends are
closed with corks glued in.   *b* is a piece
of No. 3 glass tubing 6 inches long; *c* is a piece of No.
2 of the same, drawn to a fine jet-point.   This tube must
be fitted to the cork loosely enough to allow of its being
drawn back and forth for adjustment.   At *d* the tube is
bored for the admission of a cork, through which is passed
a straight piece of No. 4 tubing 3 inches in length.   A
steady blast from the mouth through *c* will produce as
good draught through *d*, which can be connected with any
piece of apparatus through which a draught is desired.
When convenient, connect *c* by rubber tubing with the
air-cock of a steam radiator.

31. Gas Generators. — A supply of generators for H, N O,
H_2 S, etc., should be kept constantly on hand, fitted up
as shown in Fig. 21.   Select horse-radish bottles with perfectly circular mouths and lips not nicked out.   The corks
must fit the bottles nicely, and the tubes the holes in the
corks.   To ascertain if all joints are tight, close the mouth
of the delivery tube, and blow through the thistle-tube.

Fig. 10 represents a self-regulating generator for H very convenient when only a small quantity of gas is needed at a time. a is a pickle-jar, b is an olive-oil bottle with the bottom cut off, and the neck passed through the cork of the jar. The neck of the bottle is fitted with a cork and delivery tube. c is a piece of rubber tubing 6 inches long; d is a pinch-cock described below. At e is a little wire basket made of a circular piece of wire netting folded up at the edges and filled with granulated zinc. To use the apparatus, remove everything from the jar and nearly fill it with a mixture of H Cl and water, 1 part to 8, or $H_2 S O_4$, and water 1 part to 12. Then open d and slowly lower b into place. Close d; gas will be generated in b, and expel the liquid. When the acid is driven below the basket of zinc, action will cease. Upon drawing out the gas the liquid will be readmitted and more gas generated. The first bottleful of gas will be unfit for use. Let the large cork fit the bottle loosely, or bore a hole in it.

32. Gas-holders. — The aspirator, Fig. 6, can be used for a gas-holder. Fill it with water. Connect a with the gas-generator, and let out the water, regulating its exit by the rate at which the gas is generated. This can easily be done, in the case of H, by watching the end of the thistle-tube under the liquid.

Large hog's or beef's bladders make excellent gas-holders. Inflate the fresh bladder. After it is dry, smear the outside with sweet-oil, to which has been added a few drops of carbolic acid; place a little oil inside, and work the bladder thoroughly in the hands. This will prevent any drying, cracking, or offensive odor. Next insert a piece of glass tubing into the neck, and wind it very securely with linen thread. Slip over the end of the tubing a piece of rubber tubing, a few inches long, upon which is placed a pinch-cock, represented in Fig. 11.

33. Gas-cock. — Obtain of a harness-maker or hardware-dealer the clasp made to slip on a horse's rope-halter. Place over the end of the screw *b* the top of a rubber eraser from the end of a lead-pencil. This when placed upon rubber tubing makes a tight gas-cock and regulator. (Without the rubber, *Fig. 11.* Fig. 11 shows a convenient screw-clamp for battery wires.)

34. Large Receptacles for Gas. — Altogether the most satisfactory gas-holders are rubber bags, made wedge-shaped, and fitted with a well-packed cock. Two surfaces of inch board of the same shape, and with each dimension a little larger than the upper and under surfaces of the bags, should be furnished. Fasten the edges of the boards corresponding to the front end of the bag together with strong iron butts. Saw out from the edges a square opening large enough to pass the brass cock through. Then fold these wooden platforms and shut the bags between the surfaces, and draw the cock through the opening. After filling the bag, place heavy weights of stone or sand upon the upper boards. These can be kept in place by cleats nailed to the upper sloping surface. Make it a rule never to remove a weight while the issuing gas is connected with a burning jet. The relief of pressure may cause a mixture of gases to be resorbed and an explosion follow. If this precaution is observed, rubber bags are perfectly safe. Two bags of a capacity of 56 gallons fitted with gas-cocks may be bought for $ 25.00, and smaller ones at proportionate prices.

But these expensive rubber bags are not at all necessary for a school, unless it is desired to produce the calcium light.

Fig. 12 represents a very cheap and convenient gas-holder. *a* and *b* are tin paint-cans discarded in great numbers at every paint-shop. The diameter of *b* is a little less than that of *a*. A wooden keg, with one head knocked out, answers well for *a*. *c* is a piece of glass tubing, bent as shown in the figure, and passed through a cork at *d*, which is fitted to a hole in the tin or wood at that point. A wire loop is soldered to *b* at *e*, and at *f* and *f* tin sockets are soldered or riveted to receive the wooden uprights *g g*. The pulleys *h h* can be obtained of a hardware-dealer. *a* is filled with water, and nearly counterpoised by *k*, a baking-powder can furnished with a wire bail and filled with sand. The inner end of the tube *c* should rise a very little above the edge of the tank *a* to insure against the entrance of any water. On the outer end of *c* is placed a piece of rubber tubing with a pinch-cock. To fill the holder, remove *k*, open the pinch-cock and press down the receiver *b* till it is filled with water. Then hang on *k* again and connect the delivery tube *c* with the generator.

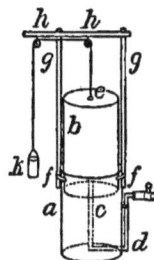

FIG. 12.

If desired, a tinman will make *a* and *b* of zinc of any required size. Make *b* 18 inches in diameter and 36 inches high, and it will hold about 40 gallons of gas. *a* should be 38 inches in height and 20 inches diameter. There should be sockets for the uprights soldered on both top and bottom of *a*; and *c* should be so placed as to issue from *a* at a point 90° from *f* and *f* instead of at the point shown in the figure. Let the wooden frame be of inch-pine stock $1\frac{1}{2}$ inches wide with the cross-bar mortised upon the uprights *g g*. If the receiver is well coated with paint it will hold O or CO_2 any length of time, but H will leak out in a very

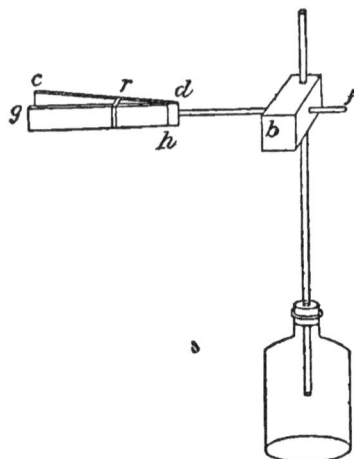

Fig. 13.

few hours. Gas cannot be subjected to great pressure in
these, on account of the overflowing of the water. The cost
of a 40-gallon gas-holder of this description is $ 5.00.

35. **Pipette-stand.** — Fig. 13 represents a pipette-stand,
useful for holding test-tubes, flasks, and small retorts while
heating. The base of the stand is a broad, low bottle filled
with sand, and the standard a piece of No. 3 glass tubing,
18 inches long; b is a piece of sheet-cork in dimensions
$2\frac{1}{2} \times 2 \times 1\frac{1}{2}$ inches. c a piece of hard wood 15 inches in
length. From c to d it is $\frac{1}{4}$ of an inch thick, $1\frac{1}{4}$ inches
wide, and 8 in length. From d to f it is turned into a
smooth rod, of the diameter of No. 3 glass tubing. g is
another piece of hard wood of the same width and thickness of c, 7 inches in length, fastened to $c f$ at h with
a brass butt. r is a stout band of rubber. It is well to
glue in thin pieces of cork flush with the inner surface

just where the necks of flasks will touch the wood. If the holes in *b* are not made too large the friction of cork and glass and wooden rods will be sufficient to retain the holder at any height and any angle.

36. Test-tube Holders. — Prepare strips of hard wood $1\frac{1}{4}$ inches wide, $\frac{1}{4}$ inch thick, and 8 inches long.   One inch from the ends hollow out the inner surface to fit the circumference of a test-tube.   Glue into these hollows pieces of asbestos-packing.   Fasten the other ends of the pieces together with brass *Fig. 14.* butts and pass a strong rubber band around them.

37. Iron Stand for Heating. — Bend 4 pieces of annealed wire of the size of telegraph-wire, 16 inches in length, in the shape represented in Fig. 14, so that the horizontal part shall be 6 inches long, and the uprights 4.   Bind the uprights together by winding them with small copper-wire in such a way as to make a square-topped stand.   Make the stand more stable by setting a large cork squarely upon the end of each leg.   Cut out of wire-netting a square 6 inches upon a side and place it upon the top of the stand.

38. Tubulated Gas-jars. — (See Fig. 15.)   These take the place of expensive bell-jars with stop-cocks, for two purposes, — for collecting gases over water and transferring to other vessels and gas-bags, and for collecting and testing combustible gases, as acetylene, etc.

For the first mentioned purpose select a large narrow-mouthed bottle, the larger the better, and cut off the bottom. (See Sec. 25.)   Through the cork, closing the mouth, pass one of the metallic-capped stoppers used in tooth-powder bottles, which can be bought of any druggist for four cents.   Settle the cork below the rim of the bottle and fill the hollow

with melted sealing-wax. To transfer gases with this, press
the cap *c* firmly upon the tube and collect the jar full of
gas over water in the ordinary way. Lift upon the jar just
enough to counterbalance its weight, slip off the cap and
put in its place the end of a rubber delivery tube. Then
sink the jar deep in the water of the pneumatic tank.

39. Sand-baths. — Obtain of a stove-dealer pieces of sheet-
iron 7 inches square. Have a square inch cut out of each
corner, and bend up a rim an inch deep on each side. If
a tight joining is not secured at the corners, fold in pieces
of asbestos-packing. Use clean beach-sand, a shallow layer.

40. Water-baths. — Cover the top of a rolled tin basin,
of a half-pint capacity, with wire-netting (mosquito-bar),

FIG. 16.

securing it firmly by a wire bound around under
the rim of the basin. Cut a large circular opening in the netting, of such a size that the evaporating-pan will settle through it two-thirds of
its depth. Cut another circular opening at one
side to receive the neck of a tall bottle. Fill the
basin to such a height with water that the bottom of the
evaporating-pan shall be submerged. Then fill the bottle
with water and invert it in the basin through the opening
made for its reception in the netting. This bottle will
keep the depth of the water constant and save the necessity
of continual watching.

41. Cork-borers. — As sets of brass cork-borers are quite
expensive, substitutes may be provided. Make cylinders
of tin 7 inches in length, and of the sizes of glass tubing
from 1 to 5. Provide a stout piece of wire of the same
length as the tubes, which will be of use to punch out the
pieces of cork from the borers, and by cutting holes 1 inch
from the top end of the larger borers and passing the wire
through horizontally, it becomes a very useful handle in
boring large corks. Keep the outer circumference of the
tubes filed to a sharp edge.

42. Deflagrating-spoons. — Make a hollow in the top of an ordinary school-crayon. Break off an inch of the crayon, and a little above the middle of this piece wind the end of a piece of copper wire 12 or 15 inches long. This can be bent at any angle for a handle.

43. Wash-bottles. — Fig. 18 (part marked b) represents a convenient arrangement for washing O, H, and other gases. It is made of a wide-mouthed quart bottle. The inlet tube reaches to the bottom of the bottle; the delivery tube b passes only through the stopper. It is well to sink the cork deep and pour in melted wax (see Sect. 58) over the top. In that case the bottle can be filled by attaching a long rubber tube to the exit tube l, placing the end of the rubber tube in a pail of water, holding the bottle below the level of the water, and sucking out a little air at the inlet k. To clean out the bottle the water can be ejected through the inlet k by blowing into the exit l. By introducing H_2SO_4 into the bottle the same piece of apparatus can be used for drying certain gases. A small hole drilled in the side of the bottle near the bottom is useful in filling the bottle and rinsing after using. It can be securely closed by means of a phial cork.

44. Drying-tubes. — They should be made of tubing at least an inch in diameter. If tubing so large is not at hand argand lamp-chimneys, either with the larger base cut off or remaining on, are excellent. If desired, soft tubing may be bent into U-tubes (see Sect. 26). But for most purposes straight tubes about 10 inches in length are very convenient. Fit corks securely to the ends, and pass through each cork pieces of No. 3 glass tubing, $2\frac{1}{2}$ inches long. Fit to the end of each tube a piece of rubber tubing 2 inches long, into which glass stoppers can be thrust to seal the tube when charged with calcium chloride, and not in use.

45. Cheap Alcohol Lamps. — Make them of wide-based mucilage or ink bottles. Fit sound corks to the necks of the bottles, and bore a little hole in the corks for a vent. Cut off two inches from a tin pea-shooter and pass it through the cork vertically. Fill this tube with strands of candle-wicking. Use spent copper pistol-cartridges for caps for the lamps.

46. Blowpipes. — Use No. 3 glass tubing. Cut off pieces 8 inches long. Two inches from an end bend the tube carefully at right angles, and draw the nearer end to a fine jet-point.

CHAPTER IV.

47. To collect Gases. — Gases may be collected in a pure state in large jars, by causing them to displace the substance already occupying the jar. What this substance should be depends upon the properties of the gas to be collected.   .

Gases insoluble in cold water, as oxygen, hydrogen, nitrogen, nitrous oxide, nitric oxide, carbonic oxide, and many of the hydrocarbon gases, may be collected in the pneumatic tank by displacement of water. Fill the gasjars completely with water. Place a cover-glass (made of common window-glass cut in squares a little larger than the diameters of the mouths of the bottles) over the mouth, and holding the cover-glass in position, with the fingers of one hand quickly invert the bottle without spilling any of the contents, and stand the inverted jar, with the mouth under water upon the shelf of the pneumatic tank or in the large earthen pneumatic pans containing two or three inches of water. Slip out the cover-glass and place three large iron nuts under the rim of the bottle's mouth in such a way as to support the bottle firmly. (A still better way is, as directed in many text-books, to file out of the rim of a flower-pot saucer a notch sufficiently large to pass the delivery tube through, and place the saucer wrong side up in the tank. Then set the jar centrally over the hole in the bottom of the saucer and push the delivery tube under it

35

through the notch.) The delivery tube of the generating
flask can then be passed under the bottles and the gas
allowed to bubble up. When the jar is full of the gas,
slip the delivery tube under the mouth of another bottle
similarly prepared. Then insert the cover-glass under the
mouth of the first bottle and set it upright upon the table.
If the gas is not to be used for some time, it is well to
smear the rim of the bottle with soft tallow before filling
it, and then to press the cover-glass down firmly before
leaving it.

Most gases soluble in water may be collected by displacement of mercury or air. As the former is expensive, and
from its great weight involves difficulties, it is not a convenient medium for use. Gases lighter than air, as ammonia, chlorhydric acid-gas, etc., are collected by downward
displacement of air. Hold the gas-bottle in an inverted
position and pass *the delivery tube upward to the bottom of
the bottle.*

Gases heavier than air, carbon dioxide, etc., are collected
by upward displacement of air. Place the gas-jar mouth
upward upon the table and pass the delivery tube down to
the bottom of the bottle.

Neither water nor mercury can be used in collecting
chlorine, for it is soluble in the one and corrodes the other.
Use hot water or strong brine, or collect it by upward displacement of air.

48. Transferring Gases from one Vessel to Another.—
Hold the jars with their mouths under water in the pneumatic tank. Insert the mouth of the vessel containing the
gas under that of the one to receive it, which must be full
of water. Then decant the first jar downward till the gas
bubbles up into the second.

**49. Two Difficulties Present Themselves in Collecting
Gas over Water.**— The pressure of the water upon the

mouth of the delivery tube causes the generator to leak, and the water is liable to be resorbed into the generating flask, when heat is employed in generating the gas.

After making the joints of the generator as tight as possible, the first trouble may be obviated by keeping the depth of water in the pneumatic pan as slight as possible, and sometimes by passing the delivery tube, at first, upward a considerable distance into the inverted jar.

The danger of resorption is much less when liquids than when dry solids are heated ; and this liability can be wholly averted in heating liquids by passing a thistle tube through the cork of the flask, the end just dipping under the liquid. With this protection, sometimes called a "safety tube," if a partial vacuum occurs, *air*, and not *water*, will enter and fill it. In heating dry substances in a glass flask, as in fact in heating anything while the mouth of the delivery tube is under water, adopt the rule *never to slacken the heat without first removing the tube from the water and shaking it free from adherent drops*.

If by any accident cold water is forced back into the hot flask, an explosion is pretty certain to result.

50. To Heat Glass Vessels without Danger of Cracking Them. — If a strong heat is needed fasten the flask securely in the retort-stand at an angle of 45° or 50°, and holding the lamp in the hand, first heat the upper part of the flask, then the lower, gently moving the flame to and fro over every part of the glass so that the expansion may be uniform. Do not let the flame play suddenly and violently upon one part. A square of wire-netting placed on the ring of the retort-stand under a heating flask or a sand-bath distributes the heat and greatly lessens the danger of breaking.

51. To Bore Corks and Fit Tubing. — Select a borer a trifle smaller than the tube to be inserted, and if necessary

enlarge the hole a very little with a round file. Always
bore through from the under surface of the cork. *Turn*
the glass tube into the cork; don't push it. Work with
the fingers close down to the cork. If the friction is likely
to be considerable, rub the tube with a piece of tallow. It
is not advisable to use hard soap for this purpose, as it
sets the tube immovably in the cork. Rubber corks may
be bored readily by keeping the borer constantly wet with
water.

52. To Evaporate Solutions. — If it is desired to obtain
crystalline salt from a solution, the more slowly the process
is carried on the larger and more perfect will be the crystals. Use a gentle heat, keeping the temperature of the
liquid just below boiling. The flame of an alcohol lamp
can be reduced by picking the wick down; of a Bunsen
lamp, by partially shutting off the gas. If this causes the
flame to snap out or descend and burn at the base, bind a
cap of fine wire-gauze over the top of the tube. Filter solutions before evaporating.

53. To Filter. — Packages of filter-paper, circular in
shape, adapted to the size of the glass funnels to be used,
6 inch paper for 3 inch funnels, should be purchased.
Fold one of these circular sheets exactly in the middle, the
crease marking a diameter, and the curved edge of the two
halves laid together. Then fold this in the middle again,
making the sheet $\frac{1}{4}$ the original size. Now open one of the
folds so that three thicknesses come upon one side and one
upon the other, and you have an inverted cone which will
just fit into the funnel. Insert the cone and place the funnel in a small ring of the retort-stand.[1] Then pour in the
liquid to be filtered and place underneath a beaker in such
a way that the lower end of the funnel shall rest against

[1] Tallow rubbed upon the under side of the lip of beakers will prevent liquids
from running down the outside in pouring.

the side of the beaker. If enough liquid is quickly poured in to nearly fill the funnel there is danger that some will work down between the paper and the glass. Avoid this by pouring a little liquid at a time down the surface of a glass rod held in nearly a vertical position with the end resting in the apex of the tunnel.

54. To Dilute Acids. — Water is miscible with the common acids in any proportion. In diluting pour the specifically heavier acid into the lighter water, *not* the water into the acid, and stir the mixture as you pour. This precaution must never be forgotten in mixing sulphuric acid and water.

55. To Remove Glass Stoppers. — If glass stoppers become fast in bottles, wrap a "bandage" of cloth wet in hot water around the neck. This must not be done with bottles containing ammonia or ether. Instead, draw a line of oil or glycerine around the edge of the stopper in the neck, and leave in a warm place for a few hours.

56. To Dry Gases. — Various methods are in use for this purpose. Allow the gas to bubble through a little strong H_2SO_4 in a wash-bottle, or fill a wash-bottle with lumps of pumice-stone wet in the acid. Another method is to pass the gas through straight tubes or bent tubes, filled loosely with calcium chloride or quicklime. Of the three substances mentioned use in each case the one which has least affinity for the gas to be dried. H_2SO_4 will soon absorb enough moisture to increase its bulk considerably, and should then be rejected for the purpose. Calcium chloride and oxide can be used several times, if the precaution is taken to plug the openings air-tight after using.

57. To Solder. — Three conditions must be fulfilled in order for success in soldering. The soldering-iron must be well "tinned." The surface to be soldered must be perfectly clean, and the novice must exercise a little patient practice in the art.

Purchase an "iron" of at least a pound's weight. With a flat file square off the faces at the point, smooth and bright, and file the sides of the iron bright and clean. Set the iron heating, and select a common brick with a slight depression near the middle of one of its large faces. Sprinkle powdered rosin abundantly into the hollow in the brick. When the iron is just below a red heat, holding a bar of solder over the brick, with the hot iron melt off a quantity, letting it drop upon the rosin. Then rub the iron vigorously against the surface of the brick through the globule of molten solder till a bright "tinned" surface appears upon the iron. In this way coat thoroughly each of the pyramidal faces at the point, and a little area back of each face. Occasionally renew this "tinning" as it becomes burnt off.

The solder may be made to adhere to new tin by merely wiping it with a cloth and sprinkling fine rosin over it. The surface of old tin must be treated with "solder fluid." Prepare this by dissolving an ounce of granulated zinc in about two fluid ounces of hydrochloric acid, adding the zinc in small successive portions, to prevent the fluid from frothing over. Do this under the hood or out of doors. A desk mucilage-bottle is convenient to keep this fluid in, and the brush belonging to such a bottle is just adapted for applying the fluid to the surface to be soldered. This chloride solution is corrosive to the surface and must be cleaned off scrupulously after the soldering is done. Have a large piece of woollen cloth or cotton-waste wet with acidulated water to wipe the hot iron upon before applying it to the solder. Melt off the solder from the bar in small quantities at a time and draw the iron slowly over the surface to be covered.

58. To prepare various Substances needed in the Laboratory. *Lime-water.* — Place a lump of quicklime, 3 inches in diameter, in a tall jar containing two quarts of water. Leave it over night. In the morning stir up the sediment. This will form "milk of lime," useful in some experiments. Allow the lime to settle. Then draw off the clear liquid with a siphon. Filtration is not necessary if the sediment is not disturbed at all while drawing off the liquid. If the water is not perfectly clear it should be filtered.

Baryta Water. — Purchase barium hydrate and prepare the water in the same manner as lime-water.

Litmus Solution and Paper. — Boil litmus in water till a strong decoction is obtained. To one portion add a few drops of caustic potash solution, for blue litmus, and to the other a few drops of dilute hydrochloric or acetic acid to make the red. These solutions are useful as long as they can be kept without fermentation. Strips of unsized papers, dipped in the liquids and dried, make very convenient test-papers. Keep these in a wide-mouthed, stoppered bottle.

Purple Cabbage Solution. — Select a small cabbage of deep color. Chop it into fine pieces and place in cold water. Let the mass simmer for at least an hour over moderate heat, then boil vigorously for a few minutes and strain off the liquid. The solution may be kept unfermented a long time by putting it up in bottles, *hot*, and sealing the corks with wax. Do not put in alcohol as a preservative.

Wax for Sealing Bottles, etc. — Melt together 10 parts by weight of pitch, 10 of gum-shellac, and 4 of beeswax. Add the shellac after the pitch and wax are completely melted, and stir until the whole mass is fluid. Then cast into sticks of convenient size. Prepare moulds for casting the sticks by filling a bucket with fine, moist sand, pressing it in hard. Make the holes for the casts with a long pencil or square wooden rod of the desired size and shapes.

Granulated Zinc. — Break or cut into pieces of convenient size, to be contained in an iron ladle or large Hessian crucible, a quantity of sheet zinc or spelter. Melt the zinc over a coal fire. With an iron spoon skim off the dross, and pour the molten metal slowly in a fine stream from a height of four or five feet into a pail of water.

59. Accidents and Emergencies in the Laboratory. — It is the duty of teachers to take every precaution with regard to the health, safety, and comfort of pupils working in a chemical laboratory. But foresight and watchfulness cannot forestall all accidents among a large number of young and inexperienced workers.

Burning Clothes and Person. — Every pupil should wear a large, strong apron of some pattern in the laboratory. Many teachers forbid the use of aprons of cotton goods on account of the ready inflammability of the material. The hot nonluminous and inconspicuous flames of the Bunsen and alcohol lamps are a constant menace to the clothing of careless pupils. Therefore recommend, if not enforce, the wearing of aprons of some cheap strong, woollen material. Keep a large old overcoat or woollen blanket hanging in the room. This wrapped closely and promptly about burning clothes will usually quench the fire without serious results. It is much more effective, and certainly preferable to dashing volumes of cold water upon a pupil.

Breathing Chlorine Gas. — With all precautions, chlorine gas is occasionally inhaled, causing violent and prolonged fits of coughing. Relief is obtained by cautiously inhaling ammonia.

Burns from Acids. — Strong acids, especially when hot, make very painful wounds in the flesh. The hot vapors of some acids like hydrobromic and bromic acids will act upon the skin producing sores. Wash such wounds immediately in a dilute solution of ammonia, caustic soda or potash, and bind up the injured parts in cloths wet in a solution of sodium carbonate.

Burns from Phosphorus. — The wounds are deep, painful, and hard to heal. Bathe the wound in carbonate of soda solution, then bind about it cotton-batting or lint soaked in the same solution, till the pain subsides. Then bandage the wound, binding upon it bits of lint soaked in glycerine diluted with one volume of water. Spirits of turpentine are sometimes used.

Burns and Scalds. — Reduce the pain and heat immediately by the free use of ice. If this is not procurable, bind on the pulp of a raw potato, finely scraped, and renew it frequently. If the skin is not abraded, great relief is obtained by covering the injured surface deeply with soft-soap. After the pain has somewhat subsided, wrap up the wound carefully, and soak the bandage with olive-oil, or, better, the well-known family remedy, "Pond's Extract of Witch Hazel."

Action of Alkalies on the Skin. — The smarting sensation caused by the caustic action of aqua ammonia and like substances is relieved by washing the skin with carbonate of soda solution.

Acid Stains on Clothing may be made less conspicuous, and sometimes obliterated by occasional applications of aqua ammonia or a solution of caustic soda or potash.

Nitric Acid Stains upon the Skin may be reduced somewhat, but not wholly effaced, by rubbing spirits of turpentine upon them. In a few days the stained cuticle may be scraped off.

CHAPTER V.

60. Hydrogen. — No materials are more satisfactory and reliable for generating H than granulated zinc and H Cl. The joints of the generator must be very tight, and the acid added in small successive quantities, not more than a teaspoonful at a time. A drop or two of a solution of platinum perchloride will sometimes make the evolution of the gas much more prompt and copious. The purity of the H must always be established before a flame is applied to it. Fill test-tubes, by displacement of water or air, with the gas and apply a burning match. As long as it burns with a shrill noise it is unsafe to use a quantity. The gas is safe when the flame passes quietly and steadily up through the tube.

61. Diffusibility of Hydrogen. — In showing the diffusion through a septum of plaster of paris in a glass tube make the plug from a sixteenth to an eighth of an inch thick, allow it to become perfectly dry before using, and avoid wetting it in the least during the experiment. After filling the tube with H, place the open end in a vessel of colored water.

The same fact is shown more simply. Place a sheet of clean blotting-paper over a jet of H, and apply a match above the paper.

62. Hydrogen Tone. — Glass tubes from one-half to one inch in diameter, and from two to three feet in length, are

right for this experiment. Either fasten the tube firmly
into the burette-holder in a vertical position and pass the
jet tube (which is inserted into the rubber delivery tube
of the generator) upward into the tube, or attach a long
glass delivery tube, terminating in a jet, in a vertical
position to the generator and pass the tube squarely down
over the flame.

63. Oxygen. — The general directions given in Chap. IV.
Sects. 47–50, in regard to heating flasks and collecting gases
over water together with the full directions usually found
in chemical text-books, will furnish sufficient aid for the
preparation of small quantities of oxygen. While the ex-
treme diffusiveness of hydrogen renders it unsafe to prepare it more than one or two hours in advance of using
it, oxygen can be kept an indefinitely long time in good
gas-holders or bladders. Oxygen should not, however, be
kept long at a time in rubber bags, as it hardens and
destroys the quality of the rubber very rapidly. It is more
convenient, and much cheaper, to prepare this gas in large
quantities, ready for use at a moment's notice. A large
glass flask may be used, but it is a gain to purchase an
iron retort, made for the purpose, with an accurately fitted
iron stopper and iron delivery tube, cost $ 3.00. Select
a retort with a well-fitted stopper *without* a clamp.

64. In Heating Large Quantities of the Mixture, the
proportions of equal weights of $K Cl O_3$ and $Mn O_2$ are by
far too wasteful. Use, instead, four parts of the chlorate
with one of the black oxide of manganese. A little more
gas may be obtained by pulverizing the chlorate crystals,
but it is safer and altogether better not to do so. Be
sure that the two substances are thoroughly mixed. Each
pound of the mixture, prepared as directed, furnishes about
thirty gallons of gas.

65. Both Oxygen and Hydrogen should be Purified by

passing them through a wash-bottle (see Sect. 43), containing for H pure water, for O either pure water or a dilute solution of caustic potash or soda.

66. The following Precautions, carefully observed, will avert all danger in preparing the gas:

First. The mixture must be absolutely free from organic matter or an explosion will result from the heating. Commercial oxide of manganese is often adulterated with charcoal or bone-black. Ascertain the purity of the materials by heating a very little of the mixture in a test-tube. If the gas passes off quietly, leaving a grayish residue, it is safe for use. (Keep the mixture in a wide-mouthed, well-stoppered bottle away from dust and impurities.)

Second. Heat the retort gently at first and lessen the intensity somewhat when the gas begins to come off abundantly.

Third. Do not let any rubber tubing that may be used become highly heated lest a compound of hydrogen, explosive with oxygen, be generated.

Fourth. Set the rubber bag or other receiver considerably higher than the level of the wash-bottle in order that no accidental rush of the gas may force water and solid particles over into the bag.

Fifth. Do not let the retort cool down at any time while connected with the wash-bottle, sufficiently to cause a resorption of water.

When the gas is all driven over, or the bag is full, slip off the tube connecting the wash-bottle and the bag from the former, and instantly close the stop-cock of the bag. Then disconnect the wash-bottle from the retort and remove the heat. When the retort is cool enough, rinse it out thoroughly with hot water, and dry it scrupulously.

67. The uses for oxygen in the laboratory are many, and the list of experiments to be performed with it is very long.

But text-books suggest so many that a considerable number in this work would be superfluous.

68. Combustibility of Oxygen. — It may be made to burn in an atmosphere of H_3N or H or illuminating-gas. In illustrating the first, select a somewhat wide-mouthed flask. Draw a piece of glass tubing to a jet-point and bend this end back in the shape of a fish-hook, so that it will readily pass down into the neck of the flask, and connect it with the delivery tube of the oxygen bag.

Place a little strong ammonia in the flask, and heat it. When the fumes begin to come from the neck abundantly, turn on the O, lower the bent tube into the neck, and apply a flame to the jet.

Fit a cork to the top of an argand lamp-chimney. Through the cork pass a delivery tube connected with a supply of H or illuminating-gas. Insert a long, straight piece of glass tubing drawn to a fine jet into the oxygen delivery tube. Fasten the chimney in a vertical position, turn on the H or illuminating-gas, ignite it at the bottom, and pass the jet-tube of O up into the chimney.

69. In Preparing Mixed Gases, place the one-third part of O in the tubulated jar *first*, and *then* run in the H. Avoid every possibility for fire to reach the mixed gases. If masses of bubbles are exploded on the surface of suds, never use a mass larger than can be covered with the two hands. The tremendous percussion does temporary and sometimes permanent injury to the "membranum tympani" of the ear. Soap-bubbles will be much more durable if to a suds made by dissolving as much hard-soap as possible in warm soft water a little glycerine is added.

70. The Principle and Working of the Compound Blow-pipe can be profitably illustrated even with bladders of two gallons capacity. If the school possesses good rubber bags of a capacity of forty gallons or more, it will pay to purchase a blowpipe, either with or without a mounting-stand, at a cost of from $ 3.50 to $ 10.00.

Fig. 17 represents a home-made substitute which is safe, and will yield very satisfactory results. *a b* is a piece of sheet-cork $2\frac{1}{2} \times 4 \times 2$ inches in dimensions. *c d* is a common mouth blowpipe. *e f* is a piece of straight brass or tin tube about the size of No. 3 glass tubing. If nothing better presents, use the larger part of another mouth blowpipe. The lime-holder *g* was in one instance made of a brass thimble with a long screw passed through the top. The thread of the screw turning in the cork served as a means for revolving and adjusting the height of the lime. A piece of fine wire-gauze should be folded over the end of the tube at *e* and securely wired to the tube. The end of the tube at *d* must be covered in the same way, and the gauze wired on very securely and neatly, so that the rubber tube will slip on over it.

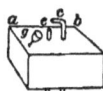
Fig. 17.

Clamp the end *b* of the block of cork firmly into the pipette or iron retort-holder. Connect *f* by means of rubber tubing with the hydrogen gas-bag. (If available, illuminating-gas will do almost as well, connecting directly upon the tube of the common burner.) Then connect *d* with the oxygen holder. First turn on the H and light it at *e*. Then slowly and cautiously turn on the O. Regulate until a slender blue flame is seen streaming straight out from the end of the oxygen tube. In this flame steel watch-springs will burn brilliantly. Small pieces of soft glass tubing and fine platinum wire will melt and drop.

71. Lime Cylinders securely packed in an air-tight tin box can be purchased for $ 1.50 per dozen. For purposes of illustration they can be prepared equally well. Select a lump of quicklime as hard and firm as possible. From it, with an old saw, cut out blocks an inch square and three inches long. Then whittle out these blocks round and smooth of the proper size to fit *g*. (Several of these cylinders can be made at once, and kept for a long time, if packed in dry sand in a fruit-jar, with the cover securely screwed on.) Let the point of the blue flame play directly upon the lime, and if it is to be used for some time, turn the lime occasionally so as to direct the flame upon a fresh spot. To obtain the best lime-light requires a little practice. Too much oxygen cools the lime, too little reduces the temperature of the flame. Too much hydrogen, especially if illuminating-gas is used, tends to crumble the lime. Remember *that the pressure upon either of the gas-holders must not be varied an ounce without shutting off both and thus extinguishing the flame.*

72. A Full Illustration of the Chemical and Physical Properties of Water and Air is important.

73. Physical Properties of Water. *Expansion when Heated.* — Fit a cork to a small test-tube and pass through the cork a piece of No. 3 or 4 glass tubing seven inches long. Make the joints very tight; fill the test-tube and the small tube up two or three inches completely (allowing no air-bubbles to become entangled) with cold water. Mark the limit of the water in the small tube by tying a thread tightly about the tube and heat it gently.

Expansion below 39.2° F. — Illustrate this by bursting a thin glass bottle. Be sure that the bottle is completely filled up to the cork, and the latter securely tied into the neck. Use a mixture of two parts of powdered ice and one of salt in a wooden vessel.

Solution of Mineral Salts in Well-water. — Demonstrate this by distilling well-water. Expensive apparatus is not needed. A glass retort and receiver even are not essential. Fit up a flask with cork and long glass delivery tube, so bent as to pass into the neck of a small flask which is to be placed in a nappie of cold water and covered with a wet towel. Evaporate very gently on a perfectly clean piece of tin or cover-glass a few drops of water from each flask. The undistilled water will leave a film of mineral salt.

The Nature of the Residue can be approximately ascertained by processes within the scope of this work.

Evaporate the water in the flask until but a few spoonfuls remain. Place portions of this residue in four test-tubes. (*a*) Add to the first half a dozen drops of HNO_3 diluted with two volumes of water, and two drops of a solution of barium chloride. Shake and hold up to the light. A dim milkiness is proof of a *sulphate* present.

(*b*) To the second add the same quantity of HNO_3 and two drops of silver nitrate. The slight cloudiness that may result will show a *chloride*.

(*c*) To the third add four or five drops of ammonia water and one or two of a solution of ammonium oxalate. If a white cloudiness appears it is evidence of a calcium salt.

(*d*) To the fourth add a few drops of H Cl and run in H_2S gas. If a very slight brownish tinge is imparted to the water a lead salt is present.

Solution of Gases. — Fill a flask and its glass and rubber delivery tube completely with water. Heat the flask, and the dissolved gases will be expelled, and can be collected over water in sufficient quantity to test.

Solution of Organic Matter. — Make a pale pink solution of potassium permanganate, and add one or two drops of sulphuric acid. Soak bits of paper in hot water for some time and pour the liquid into the permanganate solution. If the paper has been dissolved by the water in the slightest degree, the permanganate, upon standing over

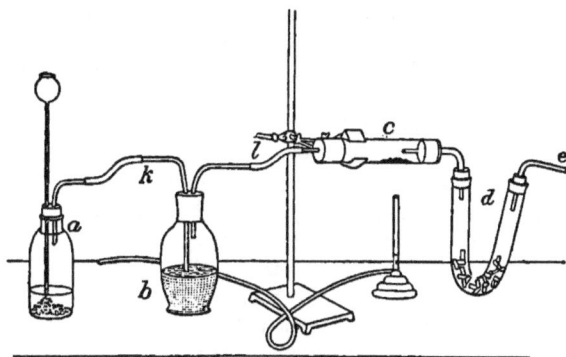

FIG. 18.

night, will give up O and lose its color. Exclude the oxygen of the air.

74. Chemical Properties of Water.

For Electrolysis of Water, see Chapter XII.

The Weight Ratios of hydrogen and oxygen in water can be demonstrated with some degree of accuracy, even before a class, by the following experiment:

a is an ordinary hydrogen generator; *b*, a small wash-bottle, one-third full of strong H_2SO_4, to absorb the moisture from the H; *c* is a tube of hard glass, one inch in diameter and twelve in length, containing a small quantity of copper oxide. Both ends of the tube are closed with close-fitting corks, through which are passed pieces of glass tubing, No. 4, three inches in length; *d* is a U-tube filled with lumps of caustic soda, or, better, calcium chloride. *Note the exact weights of the copper oxide and calcium chloride before placing them in the tubes.* It is well to weigh the calcium chloride *in the* U-*tube*, as it is somewhat difficult to remove that entirely for weighing after it has become partially dissolved in the moisture. After the apparatus is all in readiness, pour the acid into the H generator. Attach a piece of rubber tubing to e, and with it fill an inverted test-tube by displacement of air (passing the rubber tubing clear up to the bottom of the test-tube). Burn the gas thus gathered, and continue to test till the hydrogen burns quietly. Then remove the rubber from e, and with Bunsen burner or alcohol lamp heat the Cu O very cautiously at first, constantly moving the lamp to and fro. Keep the mass just below a red glow. The sides of the tube will quickly become covered with moisture from the water formed, which, as the tube becomes hotter, will all pass on into the U-tube. When the oxide is all converted into bright copper, arrest the heating, and as soon as practicable weigh the copper remaining and the calcium chloride. A deduction similar to the following may be obtained:

		GRAINS.
Weight of Cu O before heating		100
" " " after "		80
" " O in water formed		20
" " Ca Cl₂ in U-tube . . .		480
" " " after absorbing water .		. 502.5
" " water formed		22.5

(*Weight of water*) (*Weight of O*) (*Weight of H*)
22.5 grains. — 20 grains = 2.5 grains.

(*Weight of O*) (*Weight of H*)
20   :   2.5   = 8 : 1.

75. Preparation of Hydrogen Dioxide.—Barium dioxide is needed for the purpose. It may be purchased of most dealers in chemicals, or prepared with a little trouble, as directed under the subject of *Barium*. See Sect. 184.

Mix in an evaporating-pan a little pulverized dioxide with a little water to a consistency of thin cream. Then add strong H Cl in a sufficient quantity to make a quick solution. The liquid will be impure hydrogen dioxide, but will show all the properties of the substance quite well. Beautiful experiments, similar to those usually given under ozone, illustrative of its bleaching and oxidizing power, may be performed with this liquid.

76. For Synthesis and Analysis of Water, see Sects. 253 and 256.

AIR.

77. Physical Properties. *Atmospheric Pressure.* — Fit a cork closely to a thin glass flask. Fill the flask half full of water and heat it. After the water has been boiling vigorously some minutes, remove the flask from the lamp and instantly press the cork in firmly. When the flask is thoroughly cooled it will be shivered by external pressure.

Vapor of Water in Air. — Fill the bend of a U-tube with bits of caustic potash, draw a steady draught of air through the mass by connecting the tube with an aspirator-bottle, or the tube described in Sect. 30, Fig. 7. The potash will become wet, and finally dissolve in the water seized from the passing air. Or fill a tin fruit-can with a freezing mixture and note the condensation upon the outside.

Carbon Dioxide in Air. — Fill the bend of a U-tube with lime-water or baryta-water and pass a draught of air through it for a long time. Or, more simply, leave a little lime-water in a shallow vessel, like a saucer, exposed to the atmosphere for twenty-four hours.

78. Composition is usually demonstrated by shutting a measured quantity of air into a straight eudiometer tube, inserting a piece of phosphorus, and noting the decrease of volume by the oxidation of the P. Text-books direct to perform the experiment over mercury, but water may be used, if rigid accuracy is not sought after, and the experiment be finished in 10 or 12 hours. A convenient way to insert the P is to make a loop at the end of a piece of wire of proper length, melt the P under hot water, thrust the loop into the molten mass, and allow it to remain till the water is cooled. The wire can then be removed with the phosphorus adhering in the form of a pellet, and the end passed up into the confined air in the tube.

CHAPTER VI.

79. Preparation of Nitric Acid. — Use Chili saltpetre in preference to nitre, 4 parts by weight of the salt to 5 of H_2SO_4. Heat the vessel moderately, to prevent frothing.

80. Nitric Acid may be easily Decomposed with Evolution of Oxygen. — Fill a bowl of a "T D" pipe half full of HNO_3. Place the pipe in the retort-stand at such an angle that the acid will penetrate $\frac{2}{3}$ of the way through the stem. Heat the stem intensely in a Bunsen flame and pass the gas into a little bottle over water by means of a little rubber tubing slipped over the end of the stem.

81. Nitrous Oxide. — The preparation of N_2O by the decomposition of ammonium nitrate is made much more interesting by interposing between the heating flask and the receiving jar a small bottle fitted up like the wash-bottle (Chap. V. Fig. 18 *b*). Let the inside of the bottle be perfectly dry and set in a basin of cold water. The water generated in the decomposition will condense and be retained in the bottle. Make the different pieces of the delivery tube of glass with the shortest rubber connection possible, as rubber is acted upon by the laughing-gas. The same experiments in combustion can be performed with N_2O as with pure O with nearly equal brilliancy. Soap-bubbles filled with a mixture of equal parts of N_2O and H explode with deafening report.

82. Nitric Oxide. — In obtaining N O by the action of

H N O$_3$, on copper clippings, use an hydrogen generator. Cover the clippings and mouth of the thistle-tube with water. Add the acid in small successive quantities, not more than a teaspoonful at a time, and wait each time until the added portion is working before pouring in more. Continue this cautious addition till the evolution is sufficiently rapid. Phosphorus and magnesium ribbon will burn in N O, if they are allowed to get well burning before they are thrust into the gas.

83. Nitric Oxide will give up its Oxygen. — (*a*) Fill a quart jar with N O. Set the jar mouth upward on the table, covering it with a glass plate. Pour in a half teaspoonful of C S$_2$. Note the light flash as the N O gives up its oxygen to the carbon, and the sulphur of the bisulphide may be detected on the sides of the jar.

(*b*) Dissolve as large a crystal as possible of ferrous sulphate in a test-tube full of water. Add 3 or 4 drops of H$_2$ S O$_4$. Shake well and insert in water over the mouth of the delivery tube from a bottle generating N O. Fill the test-tube half full of the gas. Close the mouth of the tube with the finger, remove it from the water and shake thoroughly.

84. Nitrogen Trioxide is made by exposing N O to O of the air.

85. Nitrogen Tetroxide. — Place a little starch in a flask. Pour on it a small quantity of H N O$_3$ and warm the flask.

86. To form Nitrous Acid run the gas into water. Test it with litmus-paper.

87. The production of Nitrogen Pentoxide by the action of Cl upon nitrate of mercury is too difficult to be included within the scope of this work.

88. Ammonia. — Bottles of aqua ammonia should always be kept in a cool place and never exposed to warm sunlight. Never heat necks of bottles containing it to remove

BRIEF INSTITUTES OF GENERAL HISTORY.

By E. BENJAMIN ANDREWS, D. D., LL.D., Professor of History in Brown University. 452 pages. Cloth. Price, $2.00. Silver, Rogers & Co., Publishers, Boston.

This volume is precisely what its name implies — "institutes." As these deal with the entire sweep of history in a single volume, they must of course be "brief." The aim of the book is not to supplant the living teacher but to make him still more living. It does not offer matter for rote recitation but " blazes through the jungle of the ages a course along which the instructor can guide his class much as he lists." The method is synthetic, articulate, progressive. The work comprises eleven chapters of generic topics: The Study of History, The Old East, The Classical Period, The Dissolution of Rome, The Mediæval Roman Empire of the West, Feudalism and the French Monarchy, Islam and the Crusades, Renaissance and Reformation, The Thirty Years' War, The French Revolution, Prussia and the New Empire; each conspicuously broken up into about twenty paragraphs, after the fashion of the best German *Lehrbücher*, and all constituting together a compact, orderly, and rounded whole. Especially to encourage and facilitate collateral reading a select bibliography precedes each chapter and paragraph. All unimportant details are ignored, the most important treated in notes, and the *rationale* of historical movement made studiously prominent throughout. The result is a precipitate rather than an outline or compend, "to history what the spinal cord is to the nervous system or the Gulf Stream to the Atlantic." This is confidently believed to be for classroom use or as a Readers' Manual by far the best General History extant. Even instructors who use their own lectures will find it greatly saving of their own time, and in equal degree helpful to the progress of their classes to place these Institutes in the hands of their pupils, as a syllabus and as a register of the best literature for side readings.

The volume is handsomely printed on the best quality of paper, and bound in elegant cloth. It will grace the shelf of any library, and will prove a welcome book to all students and readers of history.

refractory glass stoppers, and always remove the stoppers of nearly filled bottles with the utmost caution. The gas is made by heating together 2 parts of ammonium chloride with one part of quicklime.

For ordinary experimental purposes obtain the gas by heating aqua ammonia in a flask provided with a delivery tube.

89. Combustibility of Ammonia in Oxygen. — Fig. 19 represents a piece of apparatus for the purpose. Fit to the top of an argand lamp-chimney a cork. Pass through the cork a piece of No. 4 tubing, drawn to a jet, and reaching within half an inch of the base of the chimney, and a short delivery tube b. Connect c with the flask of aqua ammonia, and b with a gas-bag of O. Fasten the chimney in an inverted position in a pipette-holder. When the heat begins to expel the H_3N abundantly through the fine jet, turn on the O, and when the chimney is full of the latter gas apply a match to the jet. A simple but satisfactory method is to insert a glass jet-tube, bent into the shape of a fish-hook, into the rubber delivery tube of a flask in which aqua ammonia is being heated, then lower the jet-tube into a jar of oxygen gas and apply a flame to the ammonia gas as it escapes.

FIG. 19.

90. For Electrolysis of Ammonia, see Sec. 255.

91. Solubility of Ammonia in Water. — Fit a cork to a wide-mouthed horse-radish bottle. Through the cork, and reaching two-thirds of the way to the bottom of the bottle, pass a piece of No. 4 tubing *not* drawn to a jet-point. Provide a basin of cold water colored with purple cabbage solution or litmus solution reddened with the least possible amount of acid. Connect the glass tube of the bottle with the rubber delivery tube of a flask containing strong aqua ammonia and fastened in the retort-stand ready for heating.

Holding the bottle in an inverted position and the cork very loosely in the neck to allow the escape of the confined air, fill the bottle with ammonia gas. Let the liquid in the flask boil vigorously. The moment when the bottle is full of gas can be judged pretty accurately by holding a piece of tumeric paper (yellow litmus) or a glass rod wet in H Cl at the mouth of the bottle. When all is ready, push the lamp from under the flask, slip the rubber tube off the glass tube of the bottle, and instantly placing the finger firmly over the end of the glass tube, press the cork *very* firmly into the bottle and lower the tube into the colored water and remove the thumb. In a short time the liquid will be seen creeping up the tube; as soon as the first drop enters the bottle a partial vacuum will be made, and the bottle will fill with water with a violent rush. If the absorption is not sufficiently prompt, tip the bottle down on one side so that a few drops of the liquid will run up the tube into the bottle. A cabbage solution will be turned green, red litmus blue. If failure follows two or three trials, rinse the bottle with cold water, wipe it out, and try again. Ammonia gas should always be collected by downward displacement of air.

92. Oxidation of Ammonia, under the subject of *Platinum*, see Sect. 237.

93. Ammonia may be Decomposed in a manner identical with that employed in the decomposition of CO_2, using metallic potassium. See Sect. 127, method *a*.

CHAPTER VII.

94. Hydrofluoric Acid. — The only compound of fluorine to be made in the laboratory is hydrofluoric acid. Its preparation and action on glass are described in Chap. II. Sect. 27.

95. Chlorine. — The preparation of Cl from H Cl with $Mn O_2$ is often unsatisfactory because the Cl gas is soon exhausted and only steam is driven over into the collecting jar. This may be in part obviated by adding successive portions of H Cl. One of the best methods of obtaining an abundant supply of the gas of good quality is the following: Mix four parts by weight of "coarse fine" salt with three of $Mn O_2$; place them in a flask and add water enough to form a thin paste; rinse the mixture around in the flask well and add two parts of $H_2 S O_4$. Insert the stopper and delivery tube and heat very moderately.

96. To burn Phosphorus in Cl, place a piece the size of a pea, well dried, in a deflagrating spoon and lower into a large jar of gas. It may be some time before the ignition takes place.

97. Combustion of Antimony. — It is not necessary to dry the Cl to burn Sb in it, if the jarful is collected when the gas is being evolved abundantly, and if the Sb is finely pulverized. It is well to sprinkle fine sand over the bottom of the jar before collecting the gas, as the particles of antimony burn into the glass.

59

98. Chlorine Water. — A stronger solution of Cl in water can be obtained when the water is quite cold, and if the gas is subjected to a little tension. This can be brought about thus : place a cork, water-tight, in the mouth of a glass retort in the place of the glass stopper. Completely fill the bowl, but not the long neck of the retort, with cold water. Place the retort wrong side up in a nappie and pass a piece of glass tubing down the length of the neck and well into the bowl. Connect this tubing with the rubber delivery tube from the Cl generator and run in the gas. The water will gradually be forced up the neck of the retort, and overflow into the nappie. The water will exert pressure enough to cause a greater quantity of gas to dissolve.

99. To show Decolorizing and Bleaching Power of Cl. — Use a decoction of logwood or cochineal. Soluble indigo is more difficult of reduction. For bleaching use pieces of bright calico, *wet.*

Write with common ink upon a piece of printed paper and place it in a moist condition in a jar of Cl.

100. Direct Union of Chlorine and Hydrogen. — (*a*) Draw a piece of glass tubing, No. 3 or 4 one foot in length, to a fine jet. Bend this end up in the shape of a fish-hook, so that the shorter arm shall be about two inches in length. Attach this to the rubber delivery tube of an hydrogen generator, and after properly testing, light the jet of H and lower it into a jar of Cl. Test for the formation of H Cl at the mouth of the jar, using blue litmus-paper.

(*b*) Fill a large bulging jar — an olive pickle-jar is excellent — with Cl, and pass down into it a fine jet of burning H, using a straight tube in place of the bent one. A peculiar deep bass hydrogen tone will be produced.

(*c*) *The Combustion of Turpentine* in Cl is more troublesome to show than is generally indicated in text-books.

But its accomplishment is assured if the following precautions are observed: The turpentine must be of good
quality and fresh, not oxidized in the least. Buy a little
of a druggist for the sake of a better quality. Heat
the turpentine very hot over the water-bath (see Sect.
40). Soak pieces of blotting-paper in the turpentine and
transfer them to the jar of gas as promptly as possible.

(d) Fill the small stoppered bell-jar (Sect. 38) with Cl
water. Test the water with litmus-paper. Then set in as
strong a light as possible for several hours. Test again
with litmus-paper for acid. Press the jar deep into water,
open the stop-cock and test the gas, which should be oxygen.

(e) Select a piece of glass tubing about a half-inch in
diameter and twenty inches long. Fit a cork to one end
and fill it within two inches with strong chlorine water.
Fill the remaining two inches with ammonia water. Place
the finger over the end and invert it in a wine-glass of
chlorine water. Note the bubbles, and after a time place
the finger under the mouth to prevent the liquid from
falling in the tube, remove the cork, and test the free
nitrogen in the top.

(f) Fill a soda-water bottle with hot water and invert
it over a pan of the same. Run in hydrogen till the bottle
is half filled. Then wrap a towel around it and run in Cl
till the bottle is filled, leaving a few drops of water in it.
Shake the bottle thoroughly, and, holding it horizontally
with the mouth averted, bring a lighted candle to the
mouth. After the explosion test the fumes in the bottle
with blue litmus-paper.

(g) A brilliant modification of this may be made upon
a sunny day. Select a pint bottle of clear white and rather
thin glass. Ascertain and mark upon the outside with a
file the point where it is just half full. Fill it as described

in the preceding, but be careful to have the volumes of the
two gases exactly equal. Arrange the towel so that the
bottle can be slipped instantly, and with no danger of
entanglement, from its folds, press in a cork firmly, carry
it out into the sunlight and fling the bottle high in air.
It will be shivered with a loud report.

101. Preparation of "Dutch Liquid." — Fill a small bell-
jar with hot water and invert over a pan of the same.
Run in chlorine till it is half filled. Fill the remaining
space with ethylene (see Sect. 123 *b*) and leave in strong
light a few minutes. The oil will gather on the sides of
the jar and the surface of the water.

HYDROCHLORIC ACID.

102. Preparation of Hydrochloric Acid Gas. — Text-books
direct to fuse the salt used, that a smaller surface may be
exposed to the action of the $H_2 S O_4$. This is not necessary
if well-dried "coarse fine" salt and $H_2 S O_4$ diluted, one
volume in four, are used. Mix them in the proportions by
weight of two parts of the latter to one of the former, and
heat in a flask with delivery tube. Collect gas by upward
displacement of air.

103. Solubility in Water. — Repeat the experiment in
Sect. 91, with these modifications: fill the fountain-bottle
with H Cl gas, holding the *mouth up*. Invert the bottle
over water colored with blue litmus.

104. Relation to Combustion. — Burn magnesium ribbon
in a jar of the gas.

105. Preparation from Aqueous Solution. — The gas somewhat contaminated with vapor of water can be obtained by
simply heating the commercial acid. The gas may be dried

by causing it to bubble through strong H_2SO_4 in a wash-bottle; or, better, fill the wash-bottle with bits of pumice-stone soaked in the acid, and pass the gas through.

106. Decomposition. — (a) By *Electrolysis*, see Sect. 258. (b) By *Sodium*. In the same way as CO_2, see Sect. 137 a. Use dry gas, and in place of c d a jet-tube turned upward. Ignite the H at d.

107. Chlorine Tetroxide. — Put a few crystals of $KClO_3$ in a wine-glass; cover with water, and add a few drops of H_2SO_4 through a thistle-tube.

Modify this by mixing a few tiny pieces of P with the $KClO_3$. They will burn under water.

108. Formation of Potassium Chlorate. — Fit a cork and small glass delivery tube bent at right angles to a piece of glass tubing a half-inch in diameter and four inches long. Place the mouth of this tube in a concentrated solution of caustic potash (make the solution as strong as possible, using warm water). Pass Cl gas into the tube, first sending it through a wash-bottle containing a little hot water.

109. Bromine. — This element is put up in ounce-bottles, for sale by all dealers. But it is pretty sure to leak after it is once opened, and corrode everything in its neighborhood. Enough can be easily prepared from potassium bromide to show the element and most of its properties.

Weigh out and mix four parts of finely pulverized KBr and three of MnO_2. Place in a test-tube, add four parts of H_2SO_4 and heat gently. The heavy vapors can be poured out into other test-tubes for use. Or a cork and delivery tube may be used in the tube. The tube must be of glass, quite short, and bent downward at right angles close to the cork, for the heavy vapor is hard to expel from the tube. If liquid Br is at hand, place a few drops in a small jar with a dropping-tube. Cover the jar loosely;

warm it a very little, and in a few minutes it is ready for use. *Avoid inhaling the vapor.*

110. Experiments. — (*a*) Bleach blue litmus-paper. (*b*) Pour some of the vapor into a test-tube containing a little water and shake to form bromine water, a powerful bleaching agent. (*c*) Drop finely powdered Sb into the vapor. (*d*) Burn phosphorus in it in the same manner as in chlorine, using a pint-jar. (*e*) Fill a small test-tube with bromine water and a few drops of CS_2 and shake. After it settles, note the clearness of water and the deep orange color of the globule at the bottom.

111. Starch Compound. — To a solution of starch add a solution of K Br, and pour in a little HNO_3 and shake. Make both solutions strong. Or add a very little bromine vapor to a solution of starch.

112. Hydrobromic Acid.   *Preparation.* — Place equal weights of K Br and H_2SO_4 in a test-tube and heat.

113. Iodine.   *Preparation.* — The Resublimed Iodine of the druggist is cheap and most excellent for all experimental purposes, but it can readily be prepared from an iodide. Place in a test-tube equal weights of K I, Mn O_2 and H_2SO_4 and heat. Hold the tube obliquely and invert over its mouth another cold tube.

114. Affinity for Phosphorus. — Place crystals of iodine and thin flakes of phosphorus in contact with each other on a piece of brick.

115. Starch Compound. — (*a*) To a dilute solution of starch add iodine dissolved in alcohol, or a solution of K I. (*b*) Drop some of the solution very dilute upon a fresh slice of raw potato. (In testing iodine with starch solution it must be remembered that the solution needs to be *fresh* and *cold.*) (*c*) To a dilute solution of starch add a solution of K I and shake well. Then pour in a few drops of HNO_3 and shake again.

116. Mercury Iodide. — Fasten a piece of large glass tubing, five or six inches long, in a burette-holder in a horizontal position. Insert a globule of Hg and a few crystals of I. Place a moderate heat under the Hg till it vaporizes. The ends of the tube may be loosely closed with corks. Perform this experiment under the hood.

117. Hydriodic Acid may be prepared contaminated with iodine vapor by heating equal weights of KI and H_2SO_4 in a test-tube.

CHAPTER VIII

118. Impure Carbon by the Distillation of Wood. — If a test-tube is fitted with cork and delivery tube, filled one-third full of excelsior or fine shavings and heated, a combustible gas may be gathered over water and impure carbon will remain in the tube. This "destructive distillation" will also yield a tar and a vapor in the tube which gives an acid test.

119. Porosity of Charcoal. — (*a*) Fasten a piece of charcoal, one-half inch in diameter, to a piece of annealed wire, one foot long. Heat the coal to redness, and covering it deeply with fine sand, let it cool. Fill a test-tube with ammonia gas by downward displacement of air. Bend the wire so that the charcoal can be passed up nearly the whole length of the tube. Insert the charcoal and invert the tube over a wine-glass full of mercury.

(*b*) Fill a test-tube, by downward displacement of air, with H_2S. Place the hand over the tube and turn it right side up. Drop in half a teaspoonful of powdered charcoal and shake thoroughly. Note the disappearance of the odor.

120. Decolorization by the Action of Charcoal. — Use a solution of laundress's blueing (dilute) or of litmus, soluble indigo, or purple cabbage decoction. Cochineal solution is still better. Macerate 2 or 3 cochineal grains in a mortar with a little water. Add this pulpy mass to 2 ounces of water and stir well. Add a teaspoonful of boneblack and stir very thoroughly. Then filter, more than once if necessary, to make the liquid perfectly clear.

121. Reducing Power of Carbon is illustrated by the usual experiment of directing the mouth blowpipe flame upon litharge on a piece of charcoal. Dry lead carbonate may be used instead of litharge.

Or heat one part powdered charcoal and ten of copper oxide in a test-tube, fitted with cork and delivery tube. Run the resulting gas into lime-water.

122. Formation of Lampblack. — Place upon a piece of sheet-iron or tin, supported upon a block of wood, a bit of tar or blotting-paper soaked in spirits of turpentine, or fill an alcohol lamp with turpentine; set on fire and invert a bell-jar over it, propping up the jar about a half-inch on one side.

123. Hydrocarbons. (*a*) *Methane.* — Made as described in all text-books by heating together sodium acetate, caustic soda, and quicklime in the proportions by weight of 1, 2, and 4 parts. Or soak small pieces of lime in vinegar and heat them. Do not collect the gas for some time, that the excess of steam may pass off.

(*b*) *Ethylene.* — Mix carefully one part by weight of alcohol with four of strong H_2SO_4, pouring the acid into the alcohol, stirring all the time. Heat the flask gently, and collect the gas over water in the stoppered receiver described in Sect. 38. To burn this heavy gas, open the stop-cock, and press the jar deep into the water in the pneumatic tank. If the gas has been collected in an ordinary wide-mouthed jar pour in water, as the gas burns, to drive it upward to the mouth.

(*c*) *Acetylene* can be very easily prepared by imperfectly burning illuminating-gas. Unscrew a Bunsen burner tube, turn on the gas, light it, and replace the tube. Then

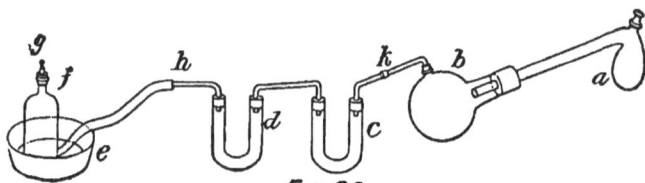

FIG. 20.

place the burner under an inverted tin tunnel. Connect the tunnel by rubber tubing with an aspirator-bottle, and draw the product, which will be impure acetylene, into the bottle. When the water has all run out, slip off the rubber tube, sink the aspirator-bottle into the pneumatic tank, light the gas at the end of the glass delivery tube, and press the bottle down into the water.

For the preparation of acetylene by means of electricity, see Sect. 257.

If illuminating-gas is not accessible, use the process by electricity, or the following:

Place in a small porcelain crucible a piece of crude tallow or lard as large as a walnut. Heat as suddenly and intensely as possible. In a short time the escaping gas can be ignited. A small Hessian crucible may be used if it is strongly heated before dropping in the lard.

124. Illuminating-Gas. — The following experiment, illustrative of the properties of illuminating-gas and of several of its waste products, may profitably be performed. *f* is a stoppered receiver of a pint capacity, made as described in Sect. 38. The gas, collected over water, is readily tested by opening the stop-cock *g*, sinking the receiver deep in the water of the pneumatic tank and applying a match. In transferring the receiver from the nappie *e* to the pneumatic tank, place a large cover-glass under the mouth of *f*. The retort *a* is one-third filled with fine soft coal, and should be heated strongly, with caution at first. The mass will swell considerably in heating. In *b,* a one-stoppered receiver, the water and tar will condense. In the U-tube *c moistened* red litmus-paper is placed to detect the presence of ammonia. In *d* are strips of paper wet in a solution of lead acetate, which will be blackened by sulphur compounds. The several joints must be made very tight.

As the experiment is quite apt to ruin *a,* it is a matter of economy to use a small retort, or substitute for it a flask or large test-tube properly fitted up. Spirits of turpentine are useful in cleaning the tar from *b* at the close of the experiment.

A similar experiment is to fit a test-tube or ignition tube with cork and glass delivery tube, drawn to a jet, and heat soft coal in it. The gas will burn, and paper moistened with lead acetate solution, and red litmus-paper, if moist, will give characteristic tests in the unlighted stream of gas.

125. Carbon Dioxide. — It is best prepared, as described in all text-books, by the action of HCl on bits of marble in a hydrogen generator. Pour in water enough to cover the end of the thistle-tube and add successive quantities of the acid, not more than a few spoonfuls at a time.

126. Properties of Carbon Dioxide. (*a*) *Its Weight.* — Fasten small " Christmas tapers " to wires two, four, and six inches long. Bore holes in a piece of board and set up the wires vertically on it, at such a distance from each other that they can be covered by an open receiver or tall pasteboard box, like a hat-box, with the bottom cut out in such a way that a narrow rim is left by which the box may be nailed to the board. Light the taper and pass the delivery tube to the bottom of the receiver or box. (In pouring CO_2 from one jar to another place the edge of the full bottle against the *outer rim* of the other. Without this precaution much of the heavy gas will tumble outside the jar.)

(*b*) *Absorption by Water.* — Fill a test-tube, whose mouth
is of the right size to be covered readily with the thumb,
with CO_2.   Collect the gas by upward displacement of air.
Introduce an inch of water.   Cover the mouth of the tube
closely with the thumb and shake well.   Place the tube in
water, remove the finger, and note the ascent of the water.
Replace the finger before raising the tube in the water, in
order to retain all the liquid that has entered.   By repeat-
ing this, all the gas may be absorbed and the tube nearly
filled with water.

(*c*) *Acid Nature when dissolved in Water.* — Run the
gas into a solution of blue litmus, or hold moistened blue
litmus-paper in a stream of the gas.

Fill a test-tube with CO_2.   Close it securely with the
·thumb, slip in a piece of caustic potash or soda, and add a
few drops of water with a dropping-tube.   Shake well (note
the heat) and invert over water as in *b*.

127. Its Decomposition.   (*a*) *By Metallic Sodium or
Potassium.* — Attach to the
generator a bulb-tube, as in
the figure.   In the bulb *b*
place a bright piece of sodium or potassium as large
as a pea.   When the gas
is passing through the tube
abundantly (which can be ascertained by running it into
a beaker of lime-water at *d*),
heat the bulb *b* moderately.
The sodium will decompose
the CO_2.   Particles of C will
be deposited in the bulb, and
may be washed from the resi-

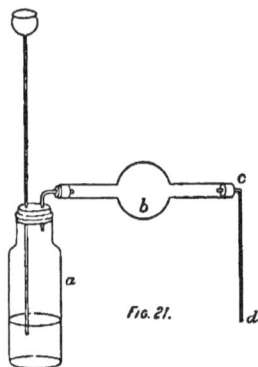

Fɪɢ. 21.

due.   The experiment will fail, if by any delay a considerable film of oxide gathers on the metal before heating.

(b) *By Magnesium.* — Fill a *large* jar with CO_2. Select a piece of magnesium ribbon long enough to reach nearly to the bottom of the jar. Make an incision in the under surface of a cork which fits the jar, and insert the end of the ribbon. Get the ribbon well to burning, lower it into the jar and press in the cork firmly. The bits of carbon cannot be washed from the insoluble oxide, as in the preceding. Therefore pour in a little water, rinse it around in the jar well, and pour into an evaporating-pan. Add a few drops of H Cl, and heat a very little. A soluble chloride will be formed, and the tiny bits of carbon will float upon the clear liquid. If the combustion of the ribbon is imperfect, the larger pieces of blackened metal may be separated by pouring the residue into a conical glass, allowing it to settle, and decanting the top of the liquid into the evaporating-pan.

(c) *By the Action of the Leaves of growing Plants.* — Dissolve as much CO_2 as possible in two quarts of water by letting the gas bubble through cold water for ten or fifteen minutes. Cut the bottom from an olive-oil bottle. Cork it tightly, fill it with the water, and invert it over a small vessel of the same water. Insert a vigorous shoot of a growing plant, — marguerite and mint are among the best, — and set the apparatus in the sunlight for several hours. To test the collected gas, remove the apparatus to the pneumatic tank. Press the bottle down into the water and remove the stopper.

128. Carbon Dioxide Produced by Fermentation. — By adding fresh yeast to a sweet solution of molasses or syrup-and-water, and leaving in a warm place for 12 hours.

129. Combustion. — The different portions of a candle-flame become more prominent by placing a dark body behind the lighted candle.

(a) To draw the gas from the centre of a candle, use

a very small glass tube, No. 7, five inches long, *not* drawn to a jet. Fasten the tube in the retort-stand at an angle of 30° from a vertical, and adjust the flame to the lower end of the tube.

(*b*) Or prepare a very small aspirator-bottle, not larger than a four-ounce, with an inlet-tube of No. 7, bent at right-angles. Insert the end of this tube in the blue cone of the flame, and let out the water slowly. To burn the gas press the bottle, with the outlet open, deep into a vessel of water, and apply a match to the end of the inlet-tube.

(*c*) Blow out a candle-flame suddenly. The uncon-sumed gas will ascend with the smoke. Hold a flame in the stream.

(*d*) To show that there is no combustion at the centre of the flame, soak a thick piece of white paper in a strong solution of alum. When it is dry press it squarely down upon the flame and suddenly remove.

Capillary Action in the Wick. — Pass a piece of old grape-vine, or, better, rattan, two inches in length through the loosely fitting cork of a small phial. Place a *very little* ether in the phial. See that the end of the rattan dips beneath the liquid, and in a moment apply a flame to the top of the rattan.

130. Proof that the Products of Combustion are Water and Carbon Dioxide. — Use a portion of the apparatus represented in Fig. 20. Place the U-tube *d*, empty and dry inside, in a beaker of cold water. Fill the bend of *c* with lime-water. Fasten a small tin funnel in an inverted posi-tion in a burette-holder, and connect the little end of the funnel by rubber tubing with *h*. Place a lighted candle under the funnel and draw the products of its combustion through the U-tubes by means of an aspirator attached at *k*. The lime-water will very quickly become milky. It will be necessary to continue the action for some time to condense a perceptible amount of water in d. For the means of obtaining a constant and steady draught, see latter part of Sect. 30. Water will always condense on the inside of a cold tumbler when it is inverted over a flame.

131. Illuminating Power of Flame Due to Carbon. — Use the apparatus represented in Fig. 21. In place of $c\ d$ use a tube drawn to a jet at d, and turned *upward*. In a generate hydrogen. In the bulb b place a little spirits of turpentine. Note the luminous nature of the flame when the gas is ignited at d. Or sprinkle charcoal-dust in an hydrogen flame.

132. Increased Weight of the Products of Combustion from Combining with Oxygen of the Air. — To the base of a large lamp chimney fit a cork. Bore several holes in the cork for the admission of air, and fasten a short piece of candle to its centre. Fit another cork to the top of the chimney.

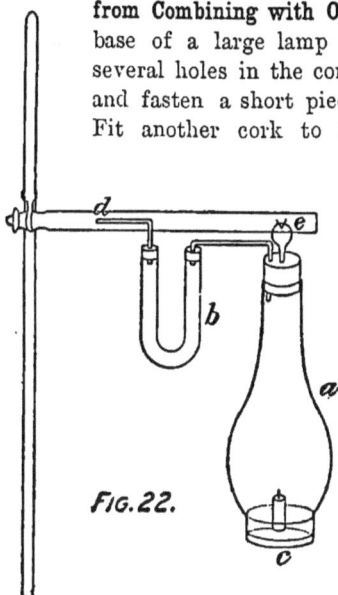

FIG. 22.

Through this cork pass a tube of No. 4 glass, and attach a four-inch U-tube, as in the figure. Turn a screw-eye e into this cork at such a point as to counterbalance the U-tube, and make the chimney hang vertically. Fill the tube b with bits of caustic-soda. Then exactly balance the whole apparatus on one arm of a delicate pair of scales. Light the candle, hang the apparatus upon a retort-stand, and with an aspirator connected at d draw the products of combustion

into *b*. The caustic-soda will retain both products. After
a few minutes remove the aspirator and return the apparatus to the scales. It will be heavier. This is an easy
experiment to make succeed if two precautions are observed. *First*, that no melted wax be allowed to drop out
of the chimney, and *second*, that a constant and steady
draught be kept up so that the waste products shall not descend upon the flame and extinguish it. To do this, make
an aspirator of several gallons capacity from a tin milk-
can, or, better, use the blast-pipe described in Sect. 30,
connecting *d* by means of rubber tubing with the side tube
a of the blast-pipe, Fig. 7.

133. Preparation of Carbon Monoxide. — Heat together
in a flask one part of powdered potassium ferrocyanide, and
ten times its weight of strong H_2SO_4. Three circum-
stances conspire to make this a troublesome experiment.
The mass swells up and froths over if heated intensely too
suddenly. Heat it gradually to the boiling-point. The gas
is apt to come off with a rush at the last. As soon as the
contents of the flask reach the boiling-point lessen the heat
a little. Finally the gas is a little soluble in water, and its
absorption may make a partial vacuum in the flask. Avoid
this danger by inserting a safety tube (see Sect. 49). The
following is a rather easier method. In a large test-tube
fitted with a cork and glass jet-tube four inches in length
place a teaspoonful of crystals of oxalic acid. Cover it
with strong H_2SO_4. Hold the test-tube at an angle of 45°
and heat moderately. When the gas begins to come off,
light the jet. Thrust the burning jet into a small-necked
bottle and note the absence of moisture. Put a little lime-
water into the bottle and shake.

If desired, a greater quantity of materials may be used,
a delivery tube substituted for the jet-tube, and the gas
collected over water. The gas will be contaminated with

some of the dioxide, which can be removed by running the gas through a wash-bottle containing a solution of caustic soda.

134. Formation of the Dioxide from the Monoxide. — Fill a jar over water two-thirds full of C O and the remaining space with O. Turn it mouth upward and apply a flame. A slight explosion will occur.

135. Preparation of Cyanogen. — Heat some crystals of yellow prussiate of potash till they are dry. Mix equal weights of this and corrosive sublimate, and place in a test-tube fitted up as in Sect. 133, last part, and heat. Ignite the gas as soon as it begins to come off. Perform all experiments with carbon monoxide and cyanogen under the hood, and clean the test-tubes used with great caution, as the gases and residues are exceedingly poisonous.

CHAPTER IX.

SULPHUR, PHOSPHORUS, ARSENIC, ANTIMONY, BORON, SILICON, AND THEIR COMPOUNDS.

136. To form Octohedral Crystals. — Upon about half a teaspoonful of flowers of sulphur in a test-tube pour twice its volume of CS_2. Cork the tube and shake occasionally. The S will not all dissolve, as a portion is likely to be of a variety insoluble in the disulphide. Filter the liquid as quickly as possible through a single thickness of perfectly dry filter-paper and expose in a shallow dish for evaporation.

137. Prismatic Crystals. — Heat bits of roll sulphur, not flowers, in a clay pipe-bowl. Heat gently just to the melting-point of the entire mass. Allow a thin crust to form in cooling, then break the crust and pour off the liquid below. If a large quantity of sulphur is melted it gives much better results. Use a very small flower-pot, previously closing the hole in the bottom with a plug of plaster-paris, and allowing the plug to dry thoroughly.

138. Plastic Sulphur. — Prepare it as directed in all text-books. Use a large test-tube, and place a cork loosely in the mouth of the tube to keep out the air and prevent the sulphur from taking fire. The contents are ready to pour after they have passed from the viscid to a thin liquid state. Pour in a fine stream into cold water and skim the residue from the surface of the water before taking out the plastic mass, which will have sunk to the bottom.

76

139. Preparation of Sulphuretted Hydrogen. — It is prepared in a manner identical with that of hydrogen. Use H Cl in preference to any other acid. The same precautions as for H must be observed in kindling the gas. The same lumps of iron sulphide may be used repeatedly, if the generator is taken to pieces immediately after using and the acid thoroughly washed away from them with clean water.

140. Decomposition of H_2 S. — Use the apparatus shown in Fig. 21. A straight tube is equally as good as the bulb tube. Turn $c\ d$ upward. As soon as the gas is ascertained to be free from air, heat the tube b gently in the middle. S will be deposited in b, and H escape at d, and may be kindled if desired.

141. Brilliant Sulphide Solutions may be made by filling test-tubes half full of aqueous solutions of lead acetate, tartar emetic, silver nitrate, copper sulphate, arsenious acid or arsenic chloride, and zinc sulphate, and running the gas into each. If any fail to form the precipitate immediately, add two or three drops of H Cl, shake, and pass in more gas.

142. Sulphuretted Hydrogen Water is formed by running the gas into cold water. Hot water will dissolve comparatively but a small quantity.

143. Hydrogen Disulphide. — Boil together 2 grammes of slaked lime, 4 of flowers of S, and 32 cc. of water. When the liquid becomes *clear* suffer it to cool, then pour into H Cl, diluted, 2 volumes.

144. Lac Sulphuris. — When the boiling liquid in the above assumes the color of gum shellac, pour off a portion into a test-tube, and after it is cool add a little H Cl, diluted, 2 volumes. Or boil together 1 gram of slaked lime, 2 of flowers of S, and 25 cc. of water, till the brown color appears. Then cool and add the dilute acid.

145. Sulphur Dioxide. — The gas contaminated with N may be made by merely burning S in a deflagrating-spoon in a fruit-jar, covering the jar loosely. It can be obtained in a pure state by treating copper or iron-filings or mercury with H_2SO_4 (the first is best). Place the filings and strong H_2SO_4 enough to cover them in an 8-ounce flask provided with cork and delivery tube. Heat gently and cautiously, lessening the heat whenever the mass shows a tendency to swell up and froth over. The gas is evolved most freely just at the boiling-point. Collect by upward displacement of air.

146. Condensation of Sulphur Dioxide to a Liquid. — By means of a rubber connector 8 or 10 inches in length, attach a three-inch U-tube (fitted up with corks and delivery tubes bent at right angles in each arm) to the delivery tube of the flask prepared for the generation of SO_2. Set up the flask ready for heating. Place the U-tube in a conical measuring-glass and pack it round with a freezing mixture of pounded ice and coarse salt. To the limb of the U-tube, not connected with the flask, attach a piece of rubber tubing and carry the tubing into a bottle of cold water in order that the excess of gas may be dissolved instead of escaping into the room. Note the action of the liquid SO_2 when poured upon water.

147. Bleaching Action of Sulphur Dioxide. — To show the bleaching action of SO_2, get a piece of roll S well to burning on a piece of broken crockery laid on a plate. Set a vase containing red roses, pansies, or oxalis flowers, previously dipped into water, on the plate and invert a bell-jar over both burning sulphur and vase of flowers.

148. The Preparation of the Troxide is difficult for an ordinary school experiment.

149. Preparation of Sulphuric Acid. — Besides the usual text-book experiment the following simple illustration of the formation of H_2SO_4 may be given. Burn a piece of roll S in a quart jar. (The S may be made to ignite more readily by turning 3 or 4 drops of alcohol over it.) Then stir the contents of the jar about with a swab wet in HNO_3. Pour in a few drops of water and shake. The water will then give an acid test with litmus-paper, and the addition of a solution of barium chloride shows the acid to be H_2SO_4.

150. Additional Experiments with H_2SO_4. — (a) Fill a narrow test-tube one-third full of strong acid. Mark the level with a string tied about the tube and leave exposed to the air a few days.

(b) Prepare a strong syrup by dissolving sugar in water. Add an equal volume of strong acid, and stir.

(c) Write upon white paper with a glass rod dipped in the acid, diluted one volume. Dry the paper at the lamp, holding the paper as near as will be safe.

PHOSPHORUS.

151. *Precautions in regard to Handling.* — Always cut and scrape phosphorus under water in a lead pan or saucer. It cannot be handled with safety, for the warmth of the hands is likely to set it on fire. If it is burned when the surface is wet the burning fragments are thrown about much more violently. Therefore always gently wipe without any friction upon soft paper all pieces intended to be burned. Keep bottles containing phosphorus well filled with water, and in a place where the water cannot freeze. Remember that it is a violent poison when taken into the stomach in the minutest quantities. For the treatment of its burns, see Sect. 59.

152. Preparation of Red Phosphorus. — Many dealers in

chemicals do not keep red phosphorus. It can be prepared when needed. Use the apparatus shown in Fig. 21. The red phosphorus can be removed more easily if a straight tube one inch in diameter is substituted for the bulb tube. In b, or the straight tube substituted, place several small pieces of P. Let d be drawn to a jet-point to insure the exclusion of the air. Generate CO_2 in the usual manner in the generator to expel the air. When the tube is certainly clear of air, carefully heat the tube under the pieces of P. The vapors, as they escape from d, will burn spontaneously. Avoid heating sufficiently for the melted phosphorus to boil, as it will then be reconverted. Continue to evolve the CO_2 till the apparatus is cooled below the kindling-point of the element.

153. Spontaneous Combustion of Phosphorus. — To cause P to ignite spontaneously with the absorbed O in charcoal, place the P on a non-conducting surface like thick blotting-paper, and cover to about the depth of $\frac{1}{4}$ of an inch with finely pulverized bone-black. It will burst into flame in from 1 to 3 minutes.

Phosphorus dissolved in CS_2, and poured over a sheet of filter-paper, will burst into flame spontaneously as soon as the CS_2 is fairly evaporated.

154. The Phosphine Rings Experiment is quite fully described in most text-books. Make the caustic solution by dissolving 20 grammes of caustic potash in 60 cc. of water. A cheaper solution, that will do nearly as well, is made by mixing 30 grammes of air-slaked lime with 100 cc. of water. Before inserting the cork and delivery tube into the flask pour in four or five drops of ether to expel the air. The gas is evolved best when the liquid boils vigorously. Keep the end of the tube scrupulously under the water in the pan. When the rings begin to come off, all currents from windows and doors, or the movements of pupils, must be avoided. The experiment should be continued but a short time; the entire apparatus should be set outside the room or under the hood, and the room thoroughly ventilated as soon as possible, as the products are quite harmful.

155. Phosphoric Anhydride may be easily made. Thoroughly dry and heat hotter than can be conveniently held in the hand, at a furnace or coal-stove, a common plate and two-quart jar. Place a piece of P twice as large as a pea upon the plate. Ignite and invert the jar over it, leaving one side canted up for the free admission of air. The flaky white powder will gather upon the sides of the jar and in the plate.

Phosphoric Acid. — Scrape the powder into a mass and add a few drops of water. It dissolves surprisingly rapidly, and a little evaporation produces the syrupy phosphoric acid.

Note the gyratory movement when flakes of the P_2O_5 are thrown upon water.

156. Hypophosphorous Acid may be made with some difficulty. Make a solution of barium hydrate. Place it in a flask and add three or four small pieces of P and a few drops of ether. Boil for some minutes. Filter and add cautiously dilute H_2SO_4.

157. The Preparation of Trioxide and Phosphorous Acids are too troublesome to be attempted with ordinary appliances.

ARSENIC.

Do all heating of oxides and other compounds of As under the hood, as the vapors, if inhaled, sometimes produce nausea, and occasionally serious illness.

158. Arsine and Marsh's Test. — Great caution should

be exercised in preparing hydrogen arsenide, and in showing the Marsh's test, as the gas, even in an impure state, is
intensely poisonous. Generate hydrogen in the usual manner in a generator provided with a glass delivery tube, 8
or 10 inches long, bent with one right angle, and drawn to
a fine jet at the end. With the usual precautions ignite
the gas. Dissolve a little white arsenic in water, or, better,
H Cl. When the flame is burning steadily pour *not more
than three or four drops* of the arsenic solution into the
thistle-tube and rinse it down with H Cl. Do not let the
flow of gas slacken, or the flame on any account become
extinguished. When the tests have been satisfactorily
made the flame should be extinguished and the apparatus
immediately taken to pieces and thoroughly rinsed. Do
this without fail under the hood or out of doors.

Test carefully with the cold porcelain before putting in
the arsenical solution, for the zinc may contain arsenic
enough to give the test.

159. Wall-papers may be Tested for arsenic by soaking
the paper for some time in warm water acidulated with
H Cl. Add a teaspoonful of this at a time to the generator,
and at the same time additional H Cl, and apply the porcelain test.

Add to the solution of wall-paper a drop of ammonia.
A blue tint is evidence of the presence of a copper salt
with which salts of arsenic are apt to be associated. If
there are green spots in the paper, drop ammonia upon
them and note if they turn blue.

160. Silver Nitrate Test of Arsenic. — In addition to
the porcelain-mirror test an interesting test may be made
if the precaution is taken to make the delivery tube in
three parts : first, the tubing bent at right angles passing
out of the bottle ; second, a piece of rubber tubing, three
or four inches long, to form a flexible joint ; and third, a

piece of glass tubing, two inches long, drawn to a jet-point. Pass the jet-tube, still ignited, down into a dilute solution of silver nitrate in a test-tube. A black precipitate of silver will presently form.

161. Scheele's Green. — Make a solution of copper sulphate and add ammonia till the precipitate redissolves. Prepare a solution of arsenic trioxide in dilute H Cl. Pour some of the sulphate solution into a test-tube and add a little of the arsenic solution. Then add successive quantities of each solution till the proper relation of acid and alkali is established.

162. Sulphide of Arsenic. — Dissolve white arsenic in water and add H_2S. The bright yellow sulphide will be precipitated.

163. Combustion of Arsenic. — Metallic As will burn brilliantly in O. Set it on fire in a deflagrating-spoon and lower into a jar of O.

164. For decomposition of As_2O_3 by metallic copper, see Sect. 260.

ANTIMONY.

165. Antimony Chloride. — The preparation of "antimony butter" is sometimes quite troublesome. Fulfil carefully these conditions: The H Cl must be strong; the Sb pulverized very fine; the ratios of materials exact (say half a gramme or 8 grains of.antimony, and 2 ounces or 60 cc. of H Cl). Before beginning to boil add 8 drops (not more nor less to the quantity of metal and acid given) of HNO_3.

166. Antimony Mirror. — To form the antimony mirror on cold porcelain proceed in a manner identical with that employed in obtaining the arsenical mirror. Pour into the generator a few drops of a solution of tartar emetic, or a very little butter of antimony, and wash it down the tube with H Cl. Then place the porcelain in the flame. An arsenical mirror can be washed away with a solution of sodium hyposulphite; the antimony stain *cannot*. Commercial antimony is quite apt to contain traces of As, and thus confuse the test if antimony chloride is used. The gas is not as dangerous to inhale as arsine.

167. For the decomposition of salts of Sb by a weak current, see Sect. 261.

BORON.

168. Borax Beads. — In working with the "borax bead" use a small platinum-wire, about 6 inches in length. Make the surface perfectly clean. Form the loop by bending the wire around some cylinder not larger than a good-sized knitting-needle. When the bead is hot let it come in contact only with the *minutest quantity of the compound to be tested.*

169. Boracic Acid. — Crystals of boracic acid are formed by dissolving borax in boiling water nearly to saturation (10 cc. of water, 5 grammes of borax) and adding strong $H_2 S O_4$ or H Cl (cautiously with the mouth of the test-tube averted from the face) to the hot solution. As the liquid cools, the crystals will appear.

SILICON.

170. Preparation of Silicon Fluoride. — Obtain silicon, fluoride as directed in all text-books, by heating together fine sand, powdered fluorspar, and strong $H_2 S O_4$. The gas is usually passed into water. The jelly-like silicic acid will clog the delivery tube. Obviate this in either of two ways. Let the delivery tube be wholly of glass, and bent twice at right angles so that the last section shall pass straight down into the water. Then, either place the water in a conical glass and pour mercury into the bottom sufficient to cover the amount of the delivery tube, or select a piece of tubing, one-half to an inch in diameter and two inches in length, and fit a cork with a small delivery tube to it. Pass this little delivery tube through a second cork which fits the tube from the flask, and insert this cork into the delivery tube of the flask, which will thus have a large calibre where it dips into the water. The jelly, as it gathers in the tube, can then be readily scraped out with the handle of a teaspoon, and clogging prevented.

171. Dialyzer. — A convenient dialyzer is made by cutting off a round wide-mouthed pint bottle about half-way up from the bottom, and stretching a piece of hog's or beef's bladder over the opening and tying it securely around the neck. Place the liquid to be dialyzed in the bottle and suspend it in two quarts or more of pure water. Several days are necessary for the complete dialysis of silicic acid.

CHAPTER X.

172. Alkali Metals. — For the spectrum analysis of compounds of these metals, see Chap. XI.

SODIUM AND POTASSIUM.

Sodium and potassium must be kept under naphtha in very tight bottles, must never be left exposed to the air, or handled with wet fingers.

173. Decomposition of Water by Sodium. — When it is desired to collect hydrogen from water by means of sodium, roll up a piece of wire-netting, four or five inches square, around a test-tube or some cylindrical body. Bend down one end of this hollow cylinder and it constitutes a wire cage. Insert the sodium, and, nipping up the other end of the cage with a pair of pincers or crucible tongs, hold it under the mouth of the collecting bottle, filled with water, which is inverted in a basin of water.

174. Protection from flying Pieces of Metal. — When pieces of either sodium or potassium are thrown upon water, care should be taken to protect the hands and eyes from flying bits of metal, as the metallic globule is apt to explode violently at the last. The water can be placed in a wide-mouthed bottle, and a cover-glass slipped over the top as soon as the metal is dropped in. Or the hands may

86

be encased in gloves and the face averted sufficiently to avoid danger to the eyes.

Sodium will inflame spontaneously only upon *hot water*.

175. Combustion in Oxygen. — Na and K will burn brilliantly in O. Heat *very small pieces* of the metals to redness in the wire cage mentioned above, and lower into a jar of O.

176. Melting in an Atmosphere of Hydrogen. — Both metals can be melted by using the apparatus shown in Fig. 21. Generate hydrogen, and when the gas is pure and abundant, insert a piece of the metal into the bulb tube and attach it to the delivery tube of the generator. After a minute or two heat the bulb cautiously. The silvery metallic nature of sodium is beautifully shown in the molten globule. When potassium is heated, a beautiful green vapor forms, which will burn spontaneously and harmlessly at *d*, making a very brilliant experiment. The bulb tube must be perfectly dry. It is well to insert a drying tube of calcium chloride between the generator and the bulb tube.

177. Combustion in Chlorine. — Both metals burn brilliantly in Cl gas. Lower tiny bits of the metals into jars of the gas. The K will take fire in a cold condition if it is placed in the jar before a heavy film of the oxide has time to form. In the case of Na it will be necessary to heat the metal in the deflagrating-spoon. It is well to protect the hand by means of a piece of pasteboard cut out a little larger than the mouth of the jar. Thrust the handle of the spoon up through the pasteboard, so that the latter shall cover the mouth of the jar when the spoon is lowered into it.

178. Additional Experiments. — (*a*) Color water with red litmus or purple cabbage solution, and throw a piece of K upon it.

(*b*) Cover the bottom of a small-necked half-pint bottle with bromine. Drop a minute piece of K into it from a deflagrating-spoon held at arm's length.

(*c*) Place a tiny flake of iodine and a thin slice of K in contact on a piece of tin placed on a ring of the retort-stand and invert a tumbler over the chemicals. Heat the tin slightly.

AMMONIUM.

179. Ammonium Amalgam. — Place 3 or 4 cc. of sodium amalgam (for the preparation of the amalgam, see Sect. 199) in a large measuring-glass, and pour over it a strong solution of ammonium chloride made by dissolving about 12 grammes of the chloride in 40 cc. of water. The mass makes an astonishing increase in bulk. Note the rapid decomposition of the amalgam when exposed to the air.

180. Prepare Ammonium Hydrosulphide by running H_2S gas for several minutes into strong ammonia water, passing the delivery tube of the gas nearly to the bottom of a test-tube two-thirds filled with ammonia.

A few drops of this hydrosulphide added to weak soultions of sulphate of zinc, copper, and iron, and to tartar emetic, will produce beautiful precipitates.

CALCIUM, STRONTIUM, AND BARIUM.

181. Decomposition of Calcium Carbonate. — The formation of "quicklime" from the carbonate is readily shown by placing a bit of marble on a piece of charcoal and directing the blowpipe flame against it.

182. Strontium and Barium Salts in Pyrotechny. — Illustrate the use of strontium and barium salts in pyrotechny as follows: Weigh out, accurately, 2 parts of strontium nitrate, 2 of potassium chlorate, and 1 of gum shellac. Pulverize them separately and very fine, and mix them thoroughly, but very cautiously, with a horn spatula or piece of pasteboard. Prepare a second mixture in the same manner as the first, using in place of the strontium nitrate the same proportion of barium nitrate. Burn a little at a time in an iron mortar, Hessian crucible, or on a brick. The former mixture yields red fire, the latter green.

183. Blanc Fixe. — Barium nitrate treated with H_2SO_4 yields "blanc fixe."

184. Preparation of Barium Peroxide. — It is important to prepare barium dioxide if it is desired to make hydrogen dioxide, as directed in Chap. V., Sect. 75.

Pulverize 2 grammes of barium oxide and mix with about $2\frac{1}{2}$ grammes of potassium chlorate. Heat the mixture in a small crucible till the oxide burns feebly but completely in the oxygen of the chlorate. Then let the mass cool and place in a close-stoppered bottle, if not immediately used. Select such a bottle that the dioxide will fill it, and if the stopper is not very tight, melt in wax about it.

SILVER.

185. Coin Solutions. — Use old-fashioned three and five cent pieces, if they can be obtained, for making solutions of silver coins.

186. The Action of Light upon Silver Salts is readily shown by adding sodium chloride, potassium iodide, and potassium bromide to solutions of silver nitrate, and exposing the precipitates to the sunlight.

187. Illustrate the Principle of Photography as follows. Cut out a piece of white paper the size of a cover-glass. Soak the paper in a solution of silver nitrate and allow it to dry in a dark place. Cut out a piece of pasteboard of the same size and remove a rectangle from the centre in such a way that the pasteboard shall resemble a picture-frame about a quarter of an inch deep. Place this frame over the glass, and the sensitive paper under it. Then place upon the frame a transparent lantern-slide, or if that is not at hand, a piece of lace, and bind all securely into place with thread. This can then be set in the sun to "print." The "tone" can be watched from the back side of the paper.

188. For the formation of the Silver Tree, see Sect. 260.

189. For the Action of Mercury upon a Nitrate Solution, see Sect. 260.

190. Explosiveness of Nitrate with Phosphorus. — Like potassium chlorate, silver nitrate will explode with a loud detonation if wrapped in paper with a bit of phosphorus and struck upon a brick with a hammer. Use only a very minute quantity of each substance.

191. Combustion of Nitrate with Charcoal. — Finely powdered nitrate and charcoal in the same condition, mixed, five parts of the salt to one of charcoal will defla-grate quite brilliantly when set on fire, leaving metallic silver.

192. Silver Oxide. — Potassium hydrate added to a dilute nitrate solution will produce a copious precipitate of silver oxide which ammonia dissolves out, leaving the liquid surprisingly limpid.

193. Silver Mirror. — A silver nitrate solution heated in a test-tube with tartaric acid leaves a silver mirror on the side of the tube.

ZINC.

194. Combustion of Zinc. — If zinc-filings and fine potassium nitrate are thrown into an *intensely hot* crucible, the zinc will burn brilliantly, producing " philosopher's wool." Drop the substances into the crucible from a long-handled spoon or deflagrating-spoon, standing away at arm's length. •

195. For the formation of Granulated Zinc, see Sect. 58.

MAGNESIUM.

196. The Action of Magnesium in decomposing carbon dioxide was shown in Sect. 127, *b*.

ALLUMINUM.

197. Formation of a " Lake." — Prepare a strong decoction of cochineal by macerating a few grains in a mortar with a few drops of cold water, and boiling this pulpy mass in 75 cc. of water. Filter and pour into a tall measuring-glass or a large test-tube and add an equal volume of a strong solution of alum. Strong ammonia will then precipitate " purple lake," which may be separated by filtering.

MERCURY.

198. Precautions in working with Hg and its Salts. — Mercury will destroy articles of gold jewelry by forming

an amalgam with them. Rings should always be removed
when working with this element. The vapors of mercury
are poisonous, and must be scrupulously avoided. The salts
are also extremely poisonous when taken into the system.

199. **Prepare "Sodium Amalgam"** as follows: Place
5 cc. of mercury in a flask. Warm the mercury a very
little, either out of doors or under the hood, and drop in
successively about 20 pieces of sodium, none of them much
larger than a pin-head.

200. **For the Separation of the Metal from the Chloride**
by a weak current of electricity, see Sect. 260.

201. **Examination of Mercuric Iodide with the Microscope.** — Make solutions of mercuric chloride, 10 grains of
the salt in 15 cc. of water; and of potassium iodide, 12
grains of the salt in 15 cc. of water, and mix. A precipitate yellow, changing to red, will form. Warm a glass
microscope slide, with a pinch of the red salt upon it, till
the color changes to yellow. Then examine the salt, as the
slide cools, under a compound microscope, using about an
one-inch objective. Illuminate the object from above with
the concave mirror or bulls-eye condenser. It makes an
extremely beautiful object for the microscope.

202. **Explosiveness of the Oxide with P.** — The red oxide
and a bit of phosphorus wrapped in paper and placed upon
a brick will explode violently when struck a smart blow
with a hammer.

203. **Formation of Calomel.** — Add oxalic acid to a solution of mercuric chloride and leave exposed to direct sunlight for a little time. Pearly flashing scales of calomel
will separate.

TIN.   ·

204. **Convenient Forms for Use in the Laboratory.** —
Chemically pure sticks and pulverized tin can be bought,

both of which are very convenient. Bar and block tin are not pure. The most convenient salts to use are the chloride. It is better to purchase these than prepare them. They may be prepared with the exercise of considerable care as follows: •

205. Preparation of Tin Chloride. — Granulate a little tin in the same way as directed to granulate zinc (Sect. 58). Weigh out a gramme of this and place it in a 4-ounce flask; add 6 cc. of strong H Cl and 3 drops of H N O$_3$ and boil for some ten minutes. Before the metal *is all dissolved*, remove the heat, and stannous chloride will be the result.

206. To form the Perchloride, use 12 cc. of H Cl. After the metal is completely dissolved, remove the liquid to an evaporating-pan; add 1 cc. of H N O$_3$ and evaporate half the bulk.

207. Tin is easily separated from a Chloride Solution. — See Sect. 260.

208. The White Tin Dioxide is made by covering pieces of tin with strong H N O$_3$. Perform under the hood.

LEAD.

209. Lead Iodide. — Add a drop or two of a solution of potassium iodide to one of lead nitrate, and acidulate the mixture with 2 or 3 drops of H Cl. Beautiful yellow lead iodide is made.

210. For the Formation of the "Lead Tree," see Sects. 259 and 260.

211. Litharge may be readily purchased.

212. The Dioxide can be purchased or prepared from commercial "red lead" which is kept in stock at any paint store. Commercial red lead is a mixture of the protoxide and true dioxide. Pour dilute HNO_3 upon the commercial article and heat a little. The protoxide will dissolve out and the dioxide remain insoluble. It can then be separated by filtering.

213. Lead "Pyrophorus." — Make a solution of "rochelle salts," and add a solution of lead acetate till the precipitate is all formed. Filter out and dry the tartrate. Half a teaspoonful or more of this should be made. Place this in an ignition tube (a test-tube is not as good, but will do if heated so as not to melt it) and heat to a red glow as long as any fumes are driven off. Before heating, select a sound cork that is a perfect fit for the tube and have it in readiness. The instant the heat is removed, insert the cork very firmly. The success of the experiment depends upon the utter exclusion of the air. It may be well, after pressing in the cork very firmly, to cut it off squarely, press it in a little farther, and spread melted sealing-wax completely over the end. When thoroughly cold, break open the tube and pour the fine powder or "lead pyrophorous" from a considerable height into a nappie. Note the formation of litharge in the nappie.

214. Lead Chloride. — Make a very strong solution of lead acetate and add HCl. Lead chloride will precipitate in crystals.

BISMUTH.

215. "Fusible Metal" is easily prepared. Melt 2 parts by weight of bismuth in a Hessian crucible. Then add one part each of lead and tin, in small portions at a time, stirring as they melt. Select a piece of No. 4 glass tubing, 6 or 8 inches long, and close one end with a phial-cork. Place the rod in a test-tube of hot water, the closed end down, and pour into this mould the hot alloy. Then remove the tube from the water. The expansion of the cooling metals will break the mould. This metallic rod placed in boiling water will soften and partly melt down.

216. Bismuth can be Separated from the Chloride by the action of a zinc strip and by electrolysis. See Sects. 259 and 261.

217. Combustion in Chlorine. — Bismuth, finely powdered, burns spontaneously when dropped into Cl gas.

218. Chloride. — Hot strong H Cl will dissolve a little of the metal forming the chloride. Do not dilute this with water, as it will thus be decomposed.

MANGANESE.

219. "Chameleon Mineral" is prepared as follows: Mix equal weights of Mn O_2 and K Cl O_3 with one-fourth their combined weight of caustic potash. Heat intensely in a Hessian crucible till the mass is green. Dissolve in cold water. The solution, at first green, will gradually turn to a splendid purple. After a few days the solution will be filled with flakes of a beautiful color, slightly iridescent.

220. Decomposition of the Dioxide. — The dioxide heated strongly in a test-tube will yield oxygen.

The same oxide heated with H Cl sets free chlorine, one method of obtaining that element.

Heated with about two-thirds its own weight of $H_2 S O_4$ it yields oxygen.

221. Spontaneous Combustion by means of the Permanganate. — Pulverize a few crystals of potassium permanganate and cautiously add the powder to strong $H_2 S O_4$. Make a pipette of a piece of No. 2 or 3 glass tubing, about

2 feet long, by drawing one end out to a moderately fine jet-point. Pour into an evaporating-pan a few drops of alcohol, ether, or bisulphide of carbon (the first is the best), and with the pipette, held in almost an horizontal position, insert into the pan of alcohol two or three drops of the permanganate solution. The alcohol will ignite spontaneously. Some volatile substances burn thus quite explosively.

IRON.

222. The Precipitates with Iron Salts are very interesting.

To a solution of caustic potash add a few drops of ferrous sulphate solution.

To a solution of potassium ferrocyanide add the iron sulphate solution. Repeat the experiment with ferricyanide.

Leave the precipitates of the first two exposed to the air for a time. The first will be reddish-brown, the second and third blue.

223. Electricity will Separate Iron from its salts. See Sect. 259.

224. Iron Hydroxide is a useful substance in case of arsenical poisoning and for other purposes. Place some clean iron-filings in a broad shallow dish as an evaporating-pan. Pour over them H_2SO_4 diluted with 7 or 8 volumes of water. Use as much as 12 cc. of the dilute acid. Cover the pan loosely with a piece of glass to prevent too rapid evaporation, and leave over night. Then filter the liquid into a test-tube and add H_3N till the hydroxide is precipitated, which can be separated by filtering. To be useful as an antidote in arsenical poisoning it must be freshly precipitated and moist. The hydroxide can be prepared in a shorter time using stronger acid, but care must then be exercised to prevent excessive frothing when the acid is poured upon the filings.

225. Sesquioxide. — Iron sulphate, when heated intensely in an ignition tube or crucible, is converted into the sesquioxide.

COPPER.

226. Nature of Salts. — The salts of copper are poisonous when taken into the system.

227. Copper Dioxide. — Place a few drops of honey in a solution of copper sulphate and add caustic soda solution till the precipitate at first formed entirely redissolves. On standing, the red dioxide will precipitate.

228. The Hydrated Oxide is obtained by adding an excess of caustic potash solution to a sulphate solution.

229. To Form the Black Protoxide, boil in separate test-tubes a rather dilute solution of copper sulphate and strong caustic soda, and pour the contents of one tube into the other while both are boiling hot.

230. Protosulphide. — To a solution of the sulphate add 2 or 3 drops of $H_2 S O_4$ and run in sulphuretted gas. The protosulphide will precipitate.

231. Chloride. — Dissolve copper oxide in H Cl and evaporate the solution slowly and gently. Green needle-like crystals of copper chloride will form.

232. "Casselman's Green." — Place in an evaporating-pan some bright copper-filings and cover them with acetic acid, and leave over night. In the morning the acid will be evaporated and the bottom of the pan covered with the green acetate. Rinse out the pan with water and pour the rinsings into a test-tube, straining out the copper clippings. Add a few drops of ammonia, bring it to the boiling-point, and pour into a boiling solution of the sulphate. The result is "Casselman's Green."

233. Nitrate. — A slow evaporation of the liquid in the generator, after making nitric oxide, will yield crystals of copper nitrate.

234. For the Separation of Copper from its Salts by Electricity, see Sect. 259.

PLATINUM.

235. Melting of Platinum. — Small platinum wire will readily melt in the oxyhydrogen flame. See Sect. 70.

236. Platinum Dissolved in "Aqua Regia." — Pour over the metal just enough of the liquid to cover it. When the action ceases, pour this off into a test-tube, and add a fresh portion of aqua regia. Thus repeat the operation until the metal is entirely dissolved. For some reason the successive portions accomplish more than the entire quantity poured over the metal at once. The aqua regia is made by adding strong nitric acid to strong hydrochloric acid just at the time that it is desired to use it.

237. Show the Oxidizing Power of Platinum upon ammonia and ether in the following way: Make a spiral, about two inches long, of rather fine platinum wire by winding it around a piece of No. 6 glass tubing. Suspend this at a convenient height from the bar of a pipette-holder or a ring of a retort-stand. Place a few drops of ether in one warm test-tube, and of strong ammonia in another. Place a lighted spirit-lamp under the spiral. When the metal is in a bright glow, remove the lamp and instantly place one of the tubes under the spiral and raise it so that the glowing metal shall be in the middle of the tube. It will continue to glow till all the vapor contents of the tube are oxidized.

238. Platinum Sponge. — Dissolve platinum in aqua regia and boil till a yellow precipitate appears. Filter out this precipitate. Wrap it as closely as possible in a piece of tough paper making a compact pellet of it. Secure it by winding fine platinum wire about it. Pass a piece of larger platinum wire through a loop in the binding wire for a handle, and heat the pellet intensely till the paper is all burned away and the salt becomes a grayish mass. It makes a very convenient pellet of platinum sponge.

Platinum chloride can be purchased, and affords a little easier means of obtaining the sponge. To the liquid chloride add a strong solution of ammonium chloride. If the precipitate fails to appear, add a few drops of H Cl. If there is a tendency on the part of the precipitate to redissolve, pour in alcohol. Then filter and heat the precipitate as directed above.

Platinum sponge that has not been used for some time will need to be cleansed. Heat it gently to quite an intense heat, and allow to cool before using.

GOLD.

239. For the Electrolysis of the Chloride and for Electroplating with Gold, see Sects. 259 and 268.

240. The Green Color of the Light transmitted by gold leaf may be illustrated. Float a gold leaf upon water. Place a cover glass in the water under the metal and carefully and squarely take up the leaf on the glass. Press the gold leaf down upon the plate by blowing with the breath directly against it till the film of water is all driven out and the metal adheres firmly and smoothly to the glass. Then hold the plate up toward the light.

CHAPTER XI.

241. Kind of Spectroscope to be Used. — Accurate and extended investigation into the spectra of elements can only be made with expensive instruments. But the spectra of several substances are well exhibited by the use of simple "pocket" spectroscopes costing from ten to sixteen dollars.

The work indicated in this chapter has been done with · a " Browning's pocket spectroscope."

242. Care of the Instrument. — The spectroscope must be kept scrupulously clean. Particles of dust adhering to the lips of the narrow slit may appear to the novice as strangely arranged bands running lengthwise of the spectrum.

243. To Adjust the Instrument. — For a class experiment, clamp the spectroscope into the pipette-holder in a horizontal position, and set the holder upon the lecture table at a convenient height. Cut out a square or circular screen of thick pasteboard, six inches in diameter. Make a hole of the proper size at the centre and slip it on the end of the spectroscope. It will afford a grateful protection to the eyes from the glaring light. An alcohol lamp will vaporize most of the compounds used, but a Bunsen burner with a ring for closing the base draughts is much to be preferred. Place the burner within one or two inches

100

of the narrow slit of the spectroscope with the draught
adjusted for a luminous flame. By means of the ring on
the tube adjust the narrow slit so that a clean sharply
defined light spectrum is obtained. Then alter the flame
to a non-luminous one. If it is necessary to use an alcohol
lamp, set the light carbonaceous flame of a candle or kerosene lamp before the instrument while adjusting for a clear
spectrum, and when that is obtained substitute the alcohol
flame in exactly the same position occupied by the bright
flame. The usual method of obtaining spectra by dipping
a piece of platinum wire in some salt, and holding it in the
flame, is impracticable before a class. The spectrum generally flashing out for an instant and disappearing, the
process must be repeated a tedious number of times.

244. Tubes for Holding Solutions. — Fig. 23 represents
a convenient piece of apparatus that will give a steady
colored flame and clear spectra for a long time.

It is made of pieces of No. 3 soft glass tubing
with one inch of their length bent at right
angles and the ends closed with small corks.
Through each cork is passed a kind of wick
made as follows : Around a rather stout piece

Fig. 23.

of platinum wire, two inches in length, are passed numerous
fine fibres of asbestos of the same length as the wire and
laid parallel with it. This little bundle is then secured by
winding with very fine platinum wire. The tubes are filled
with a solution in water of the various salts to be examined.
The fluids work slowly through the fibres of the wicks by
capillary action, and when the wicks are inserted in the
flame the characteristic coloring immediately appears, and
the spectrum can be viewed at leisure. The tubes should
be labelled with the formula of the solution and scrupulously kept for use for that one substance. The upright
axis b may be fastened into a pipette-holder or attached

to a base of its own, made from a square piece of board, and adjusted to such a height as to bring each wick upon a level with the flame, and by revolving the little turn-table each solution can be easily brought into the flame. The standard b is made of a section of broom-handle, 7 inches long, and terminating in a stout wire axis at the top. a is a disk, 6 inches in diameter, cut out of half-inch pine stock. The holes should be bored a trifle larger than the diameter of the tubes. To secure the tubes in place, cut a little square slot in the side of each hole and whittle out little wedges to fit the slots, and by pressing the wedge down beside the tube they are held securely in place.

245. Spectra of Different Metals. — With the Bunsen burner the spectra of Ba, Ca, Cu, K, Li, Mn, Na, Rb, Sr, and several others can be very readily obtained. The alcohol flame will volatilize most of these satisfactorily. The chlorides are preferable to all other salts of these metals, because of their greater solubility in water and ready volatility in flame. The carbonates of Li, and one or two others, are more easily obtained than the corresponding chloride, and though not as soluble, can be used. Strontium, and several other refractory salts, will volatilize if first treated with strong H_2SO_4. The oxyhydrogen flame, described in Chapter IV., may be used for volatilizing more refractory compounds. But a pocket spectroscope has not sufficient magnifying power for the proper examination of such resulting spectra. Very refractory substances can be volatilized by keeping the electrodes of a large Ruhmkoff's coil wet with the solution of such compounds, and passing a rapid succession of sparks. At least four Bunsen cells should be connected with the coil.

246. Absorptive Spectrum. — For the exhibition of this, an easily controlled beam of sunlight from a *porte lumiere* is necessary.

Prepare a cell for the absorptive fluid as follows: Cut from thin window-glass two plates, 3 × 4 inches. Obtain at any rubber works two or three pieces of sheet rubber, from one-fourth to one-half of an inch thick, and 3 × 4 inches in area, at a cost of 25 to 40 cents apiece. Cut out the centre of these rectangular pieces so as to form figures shaped like a in Fig. 24. These serve for the bottom and ends of the cell. At a brass foundry have four clamps cast of the shape of b for 10 or 15 cents each. The space between the prongs should be an inch, and the thread of the screw nearly as long. By means of these clamps, two at the bottom and two at the sides, fasten the rubber strips between the glass sides. Place slices of cork between the ends of the screw and the glass. Cells so made have a great advantage over all others in that they can be taken to pieces and cleaned. Cells can be made by fastening the sides to bottoms and ends of strips of glass with marine glue, but they are apt to leak and are troublesome to clean. They may be filled with aqueous solutions of blood or a decoction of logwood or cochineal. The strength of the solutions and the thickness of the strata of fluids, through which the light is passed, will be found to affect the spectra. The spectroscope should be placed close up to the cell and directly in the line of the emerging beam.

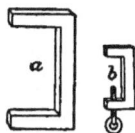

CHAPTER XII.

ELECTRICITY IN CHEMICAL REACTIONS.

247. Use of Electric Current. — Many of the most important analyses and syntheses of compounds are made, and many of the most beautiful and instructive experiments in the whole range of the study are performed by the use of the electric current and spark.

248. The Apparatus necessary consists of the various eudiometers and decomposing cells described further on, an induction coil with a $\frac{1}{4}$-inch spark, and a Bunsen battery of four cells. The cost of the coil will be $\$8.50$, of the battery about $\$2.00$ per cell.

249. Use and Care of the Battery. — It is best to avoid the nitrous fumes from the action of the Bunsen battery by using bichromate of potash solution in place of nitric acid in the porous cups. Prepare the solution by dissolving 3 ounces of bichromate of potash in two quarts of water. The outer fluid is made by mixing one part of strong H_2SO_4 with ten of water. Use the "in series" connection, a carbon with the next zinc, etc. Keep all the metallic connections rubbed bright with enamel cloth. Never leave the battery standing after use. Take it to pieces immediately. Pour the liquid from the porous cups and rinse them. Remove the zincs and rinse them thoroughly in clean water.

250. To Amalgamate the Zincs. — The zincs when purchased will be coated with mercury. This must be fre-

104

quently renewed to prevent the wear of the zincs and loss of potential by "coasting trade." Weigh out two ounces of mercury and add it to a mixture of four fluid ounces of H N O₃ and eight ounces of H Cl, and place it under the hood. After the chemical action has ceased add 10 ounces more of H Cl, and place the liquid in one of the glass jars of the battery. Dip the zincs into the liquid as deep as you desire to coat them. Do this work also under the hood or out of doors to avoid the corrosive nitrous fumes. If the zincs are small a less quantity of the liquid may be prepared, but it is economy to prepare in the proportions given. The liquid may be kept in a stoppered bottle and used for the same purpose more than once. After the zincs are once amalgamated a few drops of mercury occasionally poured directly into the outer solution, while the battery is in action, will unite with the zinc and keep the coating intact for a long time. Both fluids weaken rapidly, and therefore should be frequently renewed.

VOLUMETRIC ANALYSIS BY ELECTRICITY.

251. Decomposing Cell. — Fig. 25 represents a cell for the electrolysis of water. Provide a small tumbler and two 6-inch test-tubes of exactly the same volume. Determine this equality by filling one with water and pouring into the other. Secure the tubes in position by looping a small wire around them and binding the ends down over the edge of the tumbler. For electrodes cut out pieces of platinum foil, ¼ by 1 inch. Fasten to these platinum wires about 3 inches in length, drill the tumbler for the admission of phial-corks and pass the wire through them as shown in the figure. It is not necessary to solder the

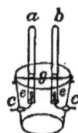

wires and electrodes. Make little slits in the foil. Pass
the wires through and head them down. This cell is
inconvenient for the electrolysis of corrosive fluids, as it
necessitates wetting the hands in testing the gases produced. A cheap and admirable cell for many experiments
is shown in figure 26.

252. Universal Electrolyzer. — Cut off from about ½-inch
glass tubing two pieces exactly 7 inches in length. Fit

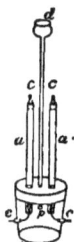

FIG. 26.

sound corks in one end of each tube. Obtain,
at any druggists, two of the patent metallic stoppers used on toilet bottles for bay-rum, tooth-
powder (select perfect ones), etc., and screw them
into the corks. Dip the corks in melted paraffine and sink them into the tubes a little more
than " flush." Fill in the space thus made above
the corks with melted sealing-wax. Select a
good tumbler of clear glass. From sheet-cork
cut a piece to fit the tumbler. Pass the two
tubes and a thistle tube through this cork, thrusting them
about an inch beyond the under surface. Prepare electrodes and platinum wires as in the cell described above.
The larger the surface of the electrodes, the more rapid
will be the action. Fit the cork into place, sinking it ¼
of an inch below the surface of the tumbler and pour
over the top enough melted sealing-wax or gum shellac to
fill the depression.

To fill the Apparatus. — Open the metallic stopper and
pour the liquid into the thistle tube. As soon as the fluid
reaches the corks in the tube, close the stopper. *Be sure
that no acids ever touch the metallic stoppers.* In testing
the gases pour more of the liquid into the thistle tube to
produce pressure, and open the cock. At p a quarter-inch
hole is drilled and stopped with a phial cork. This is for
draining the cell after using, and to let air out while filling it.

253. Analysis of Water. — Water can be made a conductor of electricity by adding about ¼ teaspoonful of sulphuric acid to the quantity necessary to use in either of the decomposing cells described. Use a four-cell Bunsen battery and connect the battery wires with the platinum wires by screw clamps (see Chap. III., Sect. 33).

254. Analysis of Hydrochloric Acid. — The equal volumes of the gases will not become manifest for some time on account of the solubility of the Cl in the fluid. It may be necessary to open the stop-cock of the H tube and let out the gas several times before the fluid is driven down equally in both tubes. Use as concentrated acid as possible. Wash out the apparatus after the experiment with care, especially the metallic cocks. The nascent Cl may attack the platinum, but the action will be slight. Electrodes may be made of pieces of small carbon pencils, but are not necessary to success.

255. Electrolysis of Ammonia. — Use the same cell in the same manner as in the preceding. Four Bunsen cells will do the work slowly. Another ready method for analyzing H_3N is as follows: Use the apparatus in Fig. 29. Pass one of the platinum wires through a cork which fits into the mouth of one of the arms of the bent tube. Fill the tube with H_3N till it touches the cork. Make the terminal which is in the arm containing the cork the negative electrode (*i. e.* connect it with the zinc), and H will gather under the cork. Connect the same electrode with the carbon, and N will gather under the cork.

256. Synthesis of Water. — It is better to purchase a Ure's eudiometer for this work, price $ 3.50. But one can be made that will do fairly well. Bend a piece of glass tubing, 3 feet long (with a half-inch bore) into the shape shown in Fig. 27. (For bending large tubing, see Sect. 26.) Seal one end in a hot flame. On opposite sides of the tube, two inches below the closed end, drill holes for the admission of phial-corks. Pass platinum wires through the corks. Coat the corks with "stratena," and press them in firmly. The wires must be exactly opposite, $\frac{1}{16}$ of an inch apart. Curve the wires into a loop outside of the corks. A barometer tube, if it does not diminish in bore toward the top, can be made into an excellent eudiometer. For making the divisions, see Sect. 27.

To use the instrument, fill the tube with water by holding it in a nearly horizontal position, the open limb uppermost, and letting the water run in a small stream into the closed limb until the bend is filled. To introduce gas, hold the tube, with mouth down, in the water. Pass the desired volume, as nearly as can be estimated, up to the bend, and invert the tube. Insert the O first. The strain of explosion is not so violent for the "home-made" instruments if an excess of one gas is used, as 4 parts of O and 4 parts of H. An excess of 2 parts of O is then left after union. Leave three inches of air in the top of the open limb for an elastic cushion, and place the thumb over the mouth.

The best apparatus for obtaining the spark is a large Ruhmkorff coil. Connect with the coil a single Grenet, or two Bunsen cells. Remove the brass electrodes and replace them with copper wire passing to the platinum loops in the eudiometer. It is well to merely lay one of the wires upon the platinum loop, that the reaction may not

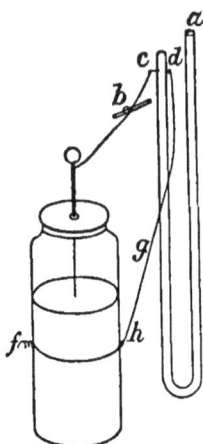

FIG. 27.

wrench out the wire from the tube. When all is ready
send a spark through the mixture. Read results only after
the water is brought to the same level in the two arms. If
the tube is not graduated, the general principle of volu-
metric union can be demonstrated by reading against a
yard-stick, placing a mark on the stick and one on the tube,
and making these marks exactly coincide at each reading.
In lieu of a Ruhmkorff coil a most excellent substitute can
be made, as in the figure, with a large Leyden jar previously
charged from a plate or Holtz machine. f is a copper wire
twisted around the outer coating of the jar, g hooks into
the loop of the wire and passes up to one of the platinum
loops. At b is a loop in the other copper wire into which
a piece of glass tubing may be inserted for a handle to
avoid the discomfort of accidentally receiving the charge.

257. To Form Acetylene. — Use the apparatus repre-
resented in Fig. 28. This consists of a tube of 1-inch bore
and 20 inches long. a is a straight champagne cork settled
in $\frac{1}{4}$ of an inch below the rim and glued in with "stratena."
The space above the cork is filled in with wax cement (see
Sect. 58). At e and b the tube is bored for the admission of
phial-corks through which are passed pieces of
carbon sticks, the size of a knitting-needle (obtained of any electrician). The electrodes should
touch. Wind the battery wires many times around
the carbon electrodes. Place several inches of
pure H in the tube over water. Pass a current
from a battery of eight cells through for some
time. Then press the tube deep into water, open
the stop-cock and ignite the acetylene.

258. To Decompose Marsh Gas, Acetylene, etc.—
Prepare a tube similar to that represented in Fig.

FIG. 28.

28, using instead of the carbon sticks platinum wires, set
$\frac{1}{16}$ of an inch apart, or use a Ure's eudiometer. Shut in a

measured quantity of the gas to be analyzed over mercury or water and pass through the apparatus a rapid succession of sparks from an induction coil. Note the change in volume of the gas and the gathering of tiny bits of C on the platinum wires.

259. Decomposition of Salts. — A description of the decomposition of a few salts will be a sufficient guide for work with as long a list as may be desired. For all such experiments use the apparatus represented in Fig. 29. The shape of the tube there given is better than that of the ordinary U-tube. Make it of a piece of tubing, 8 inches long, with a half-inch bore. The electrodes are of platinum foil, ¼ by 1 inch, and all of the wires entering the tube should be of the same metal. The number of cells to be used depends somewhat upon the results desired. For simple acid and alkali analyses two, three, or four cells may be used, the larger the number of cells the more quickly the work will be done. Make a strong solution of the salts in water.

FIG. 29.

(*a*) *Binary Salts.* — 1. Fill the tube with a solution of ammonium chloride deeply colored with a *fresh* solution of purple cabbage. Note the beautiful red caused by the acid at the + pole, and the green from the alkali at the — pole.

2. Use a clear solution of potassium bromide. Note the yellow fumes of bromine at + pole. Then use more of the same solution colored with cabbage decoction.

3. Use *first* clear solution of potassium iodide. The iodine gathers at + pole. Second, same solution with dilute cold solution of starch. Blue compound of starch forms at + pole. Third, same, colored with cabbage solution. Acid gathers at + pole, alkali at —.

4. Sodium chloride solution colored with purple cabbage.

(*b*) *Nitrates.* — Use lead, potassium, and ammonium nitrates, each colored with the cabbage solution.

(*c*) *The deposit of metals upon the — pole.* Use two Bunsen cells.

First a clear solution of iron sulphate.

Second, same of copper sulphate.

Third, of lead acetate.

Fourth, of silver nitrate. The last two, if the action is slow, will show the beautiful spangled crystallization called lead and silver "tree." To a solution of lead acetate add enough acetic acid to cause the cloudiness to disappear before passing the current through.

If the current is strong and the action rapid, the metals separate from the salt in a loose flaky powder. By a slow action copper, silver, lead, and gold may be separated so as to show their metallic properties.

260. Separation of a Metal by the Action of another Metal. — Several metals may be separated from their salts by the weak electrical action of another metal suspended in a solution of the salt.

Make an aqueous solution of lead acetate in a test tube. Dispel the cloudiness with a few drops of acetic acid and suspend a well-cleaned strip of zinc in the liquid. In this way the lead tree will form in a few hours more beautifully than by the use of a battery.

The silver tree is formed by suspending the zinc strip in a solution of the nitrate.

Silver will separate from the nitrate in an exceedingly beautiful manner by placing 2 or 3 drops of mercury in the solution. It makes the famous "Arbor Dianæ."

Separate tin from its chloride and iron from its sulphate by suspending strips of zinc in their solutions.

A strip of copper, zinc, or iron suspended in a solution of mercuric chloride sets the mercury free.

Dissolve As_2O_3 in dilute H Cl and suspend a clean copper wire in the solution. The wire will become coated with As.

261. Separation of Antimony and Bismuth by the Action between Zinc and Platinum. — Place in the bottom of an evaporating-pan a strip of platinum foil, $\frac{1}{4}$ by 1 inch (an electrode). Upon the middle of the foil place a single zinc granule so small that it will not project beyond the edge of the platinum. Pour in water enough to cover both, without disturbing either. Add H Cl till a brisk evolution of H takes place. Then add a few drops of a solution of tartar emetic for antimony, or of bismuth chloride for the latter metal. The metal will be deposited on the platinum in a black, finely divided form.

ELECTROPLATING.

.

262. General Directions. — Both electroplating and electrotyping are arts in themselves, and demand long practice and experience for the best results. But it is by no means difficult to make class illustrations of both arts, and by following directions minutely to do creditable amateur work. No expensive apparatus beyond battery cells is needed. For a plating cell cut off the neck and upper part of a junk-bottle so as to leave a square vertical-sided vessel with a capacity of about a quart. A square or rectangular shape is preferable, but a quart stone-china bowl will do. Lay stout, clean, copper wires across the vessel and suspend the anode and kathode from these by rather fine copper wire. The object to be plated is made the kathode, *i. e.* is connected with the zinc, and a piece of the same metal as that in the solution is made the anode.

263. To Clean the Articles. — Anything to be electroplated must be scrupulously cleaned, — must be *chemically* as well as physically clean. Clean off the surfaces of the metals with emery paper, crocus, or the various polishing powders in common use. Then wash the pieces thoroughly in a solution of caustic soda to remove all traces of oil or grease, and dry with a piece of soft cloth. After this stage avoid touching the surface of the metals with the hands. Next, wash the articles in dilute H_2SO_4, then in pure water, and they are ready to be quickly transferred to the plating bath.

264. Readiness of Different Metals to receive Plating. — There is a great difference in the comparative readiness with which different metals will receive metallic coatings, and in the solutions from which they will take the plating, as well.

German silver is plated most readily. Brass is a very good substance with which to practise plating. On the other hand iron and zinc cannot be plated in a copper sulphate solution with a battery. Many metals will receive the coating after being dipped in a solution prepared by dissolving mercury in nitric acid diluted with three volumes of water, and after the metal is dissolved doubling the volume by the addition of water.

265. To Copper-plate. — Use a strong solution of copper sulphate for the bath and maintain its strength with an anode of clean sheet copper of about the same area as the article being coppered. If the solution seems to weaken rapidly, add a few drops of H_2SO_4 which will go directly to the anode and cause it to dissolve more rapidly. Hang the electrodes parallel about an inch apart.

266. To Silver-plate. — The bath is a solution of silver cyanide. The solution may be purchased for about $ 2.50 per quart. This will be advantageously doubled in volume by the addition of water. (A half-pint will do a good deal of plating.) Or it may be prepared as follows: Make solutions of potassium cyanide and three times its weight of silver nitrate in equal volumes of water. When the salts are completely dissolved pour them together. After the cyanide of silver precipitate is formed, redissolve it by the addition of more potassium cyanide solution, and the clear solution is ready to be used as the bath. For want of a piece of pure silver a silver coin may be used for the anode, but after a time the copper with which the coin is alloyed works into the solution to the detriment of the latter.

267. To Nickel-plate. — For the bath purchase an ammoniacal nickel sulphate and make a weak solution in water. Or make an aqueous solution of the ordinary sulphate crystals and add aqua ammonia till the solution smells strong. Add a little more ammonia from time to time. A piece of pure nickel must be obtained for the anode.

268. For Gold Plating purchase a few grains of gold chloride. Make a solution of 1 part by weight of gold, 10 of potassium cyanide, and 100 of water. Use a gold coin for anode.

269. Plating by Chemical Replacement. — Silver may be deposited as a mirror on glass by the action of a tartrate salt on silver nitrate. To a solution of silver nitrate add weak aqua ammonia, drop by drop, till a permanent cloudiness appears. Place some of this in a clean watch-glass, add a little of a solution of " Rochelle salt," and heat the glass very gently. Place it on a sand-bath to heat. A silvered concave mirror can thus be made.

To gild on glass dip a clean glass rod into a solution of gold chloride and heat the wet rod.

Nickel plating is quite practicable by chemical replacement. Place a strong solution of zinc chloride in an evaporating-pan and dilute it with three volumes of water. If the precipitate of the salt appears, redissolve it by adding H Cl. Heat to boiling and drop in a very small pinch of zinc dust. Add a solution of nickel sulphate till the liquid is a deep green. Then drop into the boiling liquid the article of brass to be plated (cleaned with the care and in the same way as for electroplating), together with a few little pieces of zinc. Let the contents of the pan boil for ten minutes or more.

ELECTROTYPING.

270. General Directions. — The article to be electrotyped, if it is a conductor of electricity, may be hung in the bath, and after it has been covered with a sufficiently thick coat the metal stripped off, and itself hung as the kathode. The coating that then gathers will give lines of the model in relief. The practical method, however, is to take an impression of the object to be electrotyped in wax, gutta-percha, or plaster-paris, and to use this cast for the kathode. The last mentioned substance is the best for the use of the amateur because it gives the sharpest cast with the least trouble.

Select a new unworn sharp lined silver half-dollar as a good object with which to begin the art of electrotyping.

271. Preparation of the Cast. — First clean the side of the coin to be copied and rub over it a very little oil that the cast, when dry, will not adhere to it. No expensive moulding-box is needed. From a sheet of heavy writing paper cut out a strip $\frac{1}{4}$ of an inch wide and 3 or 4 inches long. Lay the coin down upon a table and measure round its circumference with the paper. Cut off the strip just a little longer than the periphery of the coin and glue the ends so that the paper loop thus made will just fit closely down over the coin. Make sure the oiled side of the coin is up and slip the paper band over the coin, making a shallow box with silver bottom and paper sides. A small rubber band stretched over the paper makes it fit the coin more closely. Make two little holes in the paper rim about $\frac{1}{18}$ of an inch apart. Select a small copper wire, 2 or 3 inches long. Pass one end of the wire in through one hole and just back through the other, and bend it so that the part of the wire within the box shall fit up close against the rim. Mix plaster of paris in water to the consistency of rather thick cream and pour it over the coin. After 2 or 3 hours the cast will be dry and can be taken out. If the cast is sharp, it may be used; if not, try again. If the copper wire was set in the box properly it will be firmly imbedded, but not completely buried, in the plaster. With a knife scrape away the plaster a little so that a clean surface of the wire can be seen the entire distance that it is imbedded.

It now remains to make the cast impervious to the water of the bath and, over the surface of the impress, a conductor of electricity. In an evaporating-pan melt some pieces of a paraffine candle, and set the reverse surface of the coin in the hot wax. The latter by capillary action will rise through the cast and spread over the entire surface. The darkening color will mark the progress of the wax on the surface.

272. Making the Cast a Conductor and Placing it in the Bath. — Cut off from a piece of pure plumbago and from a cake of "stove polish" (the stove polish alone will do, though not as well) about equal amounts. Place the two substances in a mortar and grind them together to a very fine powder. Keeping the cast warm enough over the lamp to soften the wax, with a camel's-hair brush apply this

powder to the face. Success depends upon the thorough-
ness with which this part of the work is done. *Rub and
rub the plumbago upon the face till every part and line of
the mould has a dark metallic lustre.* Carry the plumbago
surface faithfully over the edge to the wire so as to insure
good electric connection. Hang the cast as the kathode
and a clean piece of copper as the anode in a strong solution
of copper sulphate. Use one large Bunsen cell. A type
thick enough to peel off should form in 48 hours or less.

i

INDEX.

Absorption Spectrum, 102.
Acetylene, 67.
   decomposition by electricity, 109.
   formation by electricity, 109.
Acids, to dilute, 39.
Acid stains on clothing, 42.
   nitric, on skin, 43.
Air, Atmospheric Pressure, [3.
   carbon dioxide in, 53.
   composition of, 53.
   vapor of, water in, 53.
Alkalies, action of, on skin, 43.
Alkali Metals, 86.
Aluminum, 91.
Ammonia, 56.
   combustibility of, in oxygen, 57.
   decomposition of, 58.
   electrolysis of, 107.
Ammonium, 88.
   amalgam, 88.
   hydrosulphide, 88.
Ammonia, solubility of, in water, 57.
Antimony, 83.
   chloride, 83.

Antimony, combustion in chlorine, 59.
   mirror, 83.
   separated by action between zinc and platinum, 112.
Apparatus and Glassware, List of, 6, 13.
Arsenic, 81.
   combustion of, 83.
   silver nitrate test for, 82.
   sulphide of, 83.
Arsine, 81.
Aspirators, 24, 25, 26.

Barium, 88.
   peroxide, 89.
   salts in pyrotechny, 88.
Baryta-water, to prepare, 41.
Battery, care of, 104.
Bismuth, 94.
   chloride, 95.
   combustion of, in chlorine, 95.
   separated by action between zinc and platinum, 112.
" Blanc Fixe," 80.

Blowpipes, 34.
Borax Bead, 84.
Boracic acid, 84.
Boron, 84.
Bromine, 63.
   starch compound, 64.
   experiments with, 64.
Breathing chlorine gas, 42.
Burns from acids, 42.
   phosphorus, 43.
   and scalds, 43.
Burning clothes and person, 42.

Cabbage, purple solution, to prepare, 41.
Calcium, 88.
   carbonate, decomposition of, 88.
Calomel, 92.
Capillary action in wicks, 72.
Carbon by distillation of wood, 66.
   reducing power of, 67.
   dioxide, absorption of, by water, 70.
      acid nature in water, 70.
      decomposition of, 70.
      in fermentation, 71.
      from monoxide, 75.
      preparation of, 69.
      properties of, 69.
   monoxide, preparation of, 74.
Casselman's Green, 97.
Cast, preparation of, 115.
Chamelion Mineral, 95.
Charcoal, decolorizing power, 66.
   porosity of, 66.
Chemicals, List of, 7, 14.

Chlorine, 59.
   decolorizing and bleaching power of, 60.
   direct union with hydrogen, 60.
   tetroxide, 63.
   water, 68.
Combustion, 70.
   proof of products, 72.
   weight of products of, 73.
Compound blowpipe, 47.
Copper, 97.
   chloride, 97.
   dioxide, 97.
   hydrated oxide of, 97.
   nitrate, 98.
   plating, 113.
   protoxide, 97.
   protosulphide, 97.
Cork-borers, 32.
Corks, to bore, 37.
Cyanogen, 75.

Decomposing bell, 105.
Deflagrating-spoons, 33.
Dialyzer, 85.
Drying-tubes, 33.
Dutch Liquid, preparation of, 62.

Electric current, use of, 104.
Electricity, apparatus for, 104.
   in chemical reactions, 104.
Electrolyzer, universal, 106.
Electroplating, general directions, 112.
Electrotyping, general directions, 115.
Ethylene, 67
Evaporating solutions, 38.

Filtering, 38.
Fusible Metal, 94.

Gas-cock, 28.
  holders, 27.
  jars, tubulated, 31.
Gases, to collect, 35.
  difficulties of collecting, 36.
  to dry, 39.
  to transfer, 36.
Generators, 26.
Glass, bending, 21.
  boring, 18.
  to close tubes, 23.
  cutting, 19, 20.
  etching, 22.
  grinding, 19.
  stoppers, to remove, 39.
  tubing, supply of, 18.
Gold, 99.
  green color of, 99.
  plating, 114.

Heating, 3.
  glass vessels, 37.
  iron stand for, 31.
Hood, 5, 12.
Hydrobromic acid, preparation of, 64.
Hydrochloric acid, decomposition of, 62.
  electrolysis of, 107.
  preparation of, 62.
  relation to combustion, 62.
  solubility in water, 62.
Hydrofluoric acid, 59.

Hydrogen, 44.
  diffusibility of, 44.
  dioxide, preparation of, 52.
  disulphide, 77.
  to purify, 45.
  tones, 44.
Hydriodic acid, 65.
Hypophosphorus acid, 81.

Iodide of mercury, 65.
Iodine, 64.
  affinity of, for phosphorus, 64.
  preparation of, 64.
  starch compound, 64.
Illuminating-gas, 68.
  power of flame, 73.
Iron, 96.
  hydroxide, 96.
  salts, precipitates of, 96.
  sesquioxide, 97.

Labelling chemicals, 17.
Laboratory, cost of, 9, 16.
  table, 3, 10.
Laboratories, the two, 2.
Lac sulphuris, 77.
Lakes, 91.
Lamps, alcohol, 31.
Lampblack, formation of, 67.
Lead, 93.
  chloride, 94.
  dioxide, 93.
  iodide, 93.
  "pyrophorus," 94.
Lecture-table, 4, 11.
Lime cylinders, 48.
  water, to prepare, 40.
Litharge, 93.

Litmus-paper, to prepare, 41.
  solution, to prepare, 41.

Magnesium, 91.
Manganese, 95.
  dioxide, decomposition of, 95.
Marsh-gas, decomposition of, by electricity, 107.
Marsh's test, 81.
Mercury, 91.
  precaution in working with, 91.
Mercuric iodide under the microscope, 92.
  oxide with phosphorus, 92.
Metals, 86.
  readiness of, to receive plating, 113.
  separated from salts, 111.
Methane, 67.
Mixed gases, 47.

Nickel-plating, 114.
Nitric acid, decomposition of, 55.
  preparation of, 55.
  oxide, 55.
    gives up oxygen, 56.
Nitrous acid, to form, 56.
  oxide, 55.
Nitrogen pentoxide, 56.
  tetroxide, 56.
  trioxide, 56.

Oxygen, 45.
  combustibility of, 47.
  heating the mixture, 45.
  precautions in preparing, 46.
  to purify, 45.

Permanganate of potassium, spontaneous combustion of, 95.
Phosphine rings, 80.
Phosphoric acid, 81.
  anhydride, 81.
  trioxide, 81.
Phosphorous acid, 81.
Phosphorus, 79.
  combustion of, in chlorine, 59.
  precaution in handling, 79.
  red, 79.
  spontaneous combustion of, 80.
Photography, 90.
Pipette stand, 30.
Plating, by chemical replacement, 114.
Platinum, 98.
  dissolved in "aqua regia," 98.
  melting of, 98.
  oxidizing power of, 98.
  sponge, 99.
Plumbing, 13.
Potassium, 86.
  chlorate, formation of, 63.

Receptacles for gas, large, 28.
Rooms, selection of, 1.

Sand-baths, 32.
Salts, electrolysis of, 110.
Scheele's green, 83.
Sealing-wax, to prepare, 41.
Silicon, 84.
  flouride, 84.
Silver, 89.
  coin solution, 89.

Silver, mirror, 90.
   nitrate with charcoal, 90.
      with phosphorus, 90.
   oxide, 90.
   plating, 113.
   salts, action of light upon, 87.
Sodium, 86.
   amalgam, 92.
   combustion of, in chlorine, 87.
   combustion of, in oxygen, 87.
   experiments with, 87.
   melted in hydrogen, 87.
   protection from pieces of, 86.
Soldering, 39.
Spectra of different metals, 102.
Spectroscope, to adjust, 100.
   care of, 100.
   kind of, to use, 100.
Spectrum analysis, 100.
      tubes for solutions, 101.
Storage-closets, 4, 11.
Strontium, 88.
   salts in pyrotechny, 88.
Sulphide solutions, 77.
Sulphur, 76.
   dioxide, 78.
      bleaching action of, 78.
      condensation to a liquid, 78.
   plastic, 76.
   octahedral crystals of, 76.
   prismatic crystals of, 76.
   trioxide, 78.

Sulphuretted hydrogen, 77.
      decomposition of, 77.
      water, 77.
Sulphuric acid, 78.
      experiments with, 79.
Test-tube holders, 31.
Tin, 92.
   chloride, 93.
   dioxide, 93.
   perchloride, 93.
Turpentine, combustibility of, in chlorine, 60.
Wall-papers, testing of, for arsenic, 82.
Wash-bottles, 33.
Water, decomposition of, by sodium, 86.
   expansion of, below 39.2° F., 49.
   gases in, 50.
   mineral salts in, 49.
   organic matter in, 50.
   physical properties of, 49.
   supply of, 2.
   synthesis of, 107.
   weight, ratios of oxygen and hydrogen in, 51.
Water-baths, 32.
Zinc, 91.
   combustion of, 91.
   to prepare granulated, 42.
Zincs, to amalgamate, 104.